CROP PRODUCTION SCIENCE IN HORTICULTURE SERIES

Series Editors: Jeff Atherton, Senior Lecturer in Horticulture, University of Nottingham, and Alun Rees, Horticultural Consultant and Editor, *Journal of Horticultural Science*.

This series examines economically important horticultural crops selected from the major production systems in temperate, subtropical and tropical climatic areas. Systems represented range from open field and plantation sites to protected plastic and glass houses, growing rooms and laboratories. Emphasis is placed on the scientific principles underlying crop production practices rather than on providing empirical recipes for uncritical acceptance. Scientific understanding provides the key to both reasoned choice of practice and the solution of future problems.

Students and staff at universities and colleges throughout the world involved in courses in horticulture, as well as in agriculture, plant science, food science and applied biology at degree, diploma or certificate level will welcome this series as a succinct and readable source of information. The books will also be invaluable to progressive growers, advisers and end-product users requiring an authoritative, but brief, scientific introduction to particular crops or systems. Keen gardeners wishing to understand the scientific basis of recommended practices will also find the series very useful.

The authors are all internationally renowned experts with extensive experience of their subjects. Each volume follows a common format covering all aspects of production, from background physiology and breeding, to propagation and planting, through husbandry and crop protection, to harvesting, handling and storage. Selective references are included to direct the reader to further information on specific topics.

Titles Available:
1. **Ornamental Bulbs, Corms and Tubers**   A.R. Rees
2. **Citrus**   F.S. Davies and L.G. Albrigo
3. **Onions and Other Vegetable Alliums**   J.L. Brewster
4. **Ornamental Bedding Plants**   A.M. Armitage
5. **Bananas and Plantains**   J.C. Robinson
6. **Cucurbits**   R.W. Robinson and D.S. Decker-Walters
7. **Tropical Fruits**   H.Y. Nakasone and R.E. Paull
8. **Coffee, Cocoa and Tea**   K.C. Willson
9. **Lettuce, Endive and Chicory**   E.J. Ryder
10. **Carrots and Related Vegetable *Umbelliferae***   V.E. Rubatzky, C.F. Quiros and P.W. Simon
11. **Strawberries**   J.F. Hancock
12. **Peppers: Vegetable and Spice Capsicums**   P.W. Bosland and E.J. Votava

PEPPERS: VEGETABLE AND SPICE CAPSICUMS

P.W. Bosland
and
E.J. Votava

Department of Agronomy and Horticulture
New Mexico State University
Las Cruces
USA

CABI *Publishing*

CABI *Publishing* **is a division of CAB** *International*

CABI Publishing	CABI Publishing
CAB International	10 E 40th Street
Wallingford	Suite 3203
Oxon OX10 8DE	New York, NY 10016
UK	USA
Tel: +44 (0)1491 832111	Tel: +1 212 481 7018
Fax: +44 (0)1491 833508	Fax: +1 212 686 7993
Email: cabi@cabi.org	Email: cabi-nao@cabi.org
Web site: www.cabi-publishing.org	

A catalogue record for this book is available from the British Library, London, UK.

Library of Congress Cataloging-in-Publication Data
Bosland, Paul W.
    Peppers: vegetable and spice capsicums /P.W. Bosland and E. Votava.
      p.   cm.— (Crop production science in horticulture; 12)
    Includes bibliographical references (p.).
    ISBN 0-85199-335-4 (alk. paper)
    1. Peppers. I. Votava, E. (Eric) II. Title. III. Series.
SB351.P4B67   1999
635′.6437—dc21   99-16961
                                                      CIP

ISBN 0 85199 335 4

First printed 2000
Reprinted 2003

Typeset by AMA DataSet Ltd, UK.
Printed and bound in the UK by Biddles Ltd, Guildford and King's Lynn.

CONTENTS

PREFACE

Peppers, chiles, capsicum, or whatever they are called, are a versatile crop. The new fascination with capsicum peppers is reflected in seed catalogues that now advertise more types than ever before. This increased interest can be attributed to the rich diversity that the crop has to offer. The wide variety of uses include flavouring in food manufacturing, colouring for cosmetics, and imparting heat to medicines. Some are used as ornamental plants. The dried red powder is even used to colour flamingos in zoos and koi in aquariums! These 'hidden' uses of peppers are what makes *Capsicum* a major commodity, even though it is often listed as a 'minor crop'.

In writing this book, we were faced with the dilemma as to what word to use for the *Capsicum* genus. The aetiology of *Capsicum* terminology is confusing. The italicized and capitalized word *Capsicum* is reserved for taxonomic discussion, while the lower case and non-italicized word 'capsicum' is used as a vernacular term. A common and erroneous name for the genus is red pepper. This is an unfortunate misnomer because true peppers are different plants entirely, not related to *Capsicum*, in the slightest sense. True pepper belongs to the *Piperaceae* family. In addition, the *Capsicum* genus goes by an innumerable set of common names, such as pepper, chili, chile, chilli, aji and paprika. The Spanish word 'chile' is a variation of the phonetic 'chil' sound derived from the Nahuatl (Aztec) language, whereas 'aji' is a variation of 'axi' from the extinct Arawak language of the Caribbean. Because the Aztecs did not have a written language, it is impossible to know the true spelling of the word 'chile'. In the Spanish speaking parts of Mexico, Central America, and the south-western United States, *Capsicum* is called 'chile'. In the United States, a Senator from New Mexico, the Honourable Pete Domenici, put into the Congressional Record 'the correct way to spell chile' (Domenici, 1983). Today, the word 'chile' is used for the plant and the fruit, whereas the anglicized form 'chili' is used for a specific dish of food. The state of Texas in the United States has an official state dish made with beans, meat and chile and this food is spelled chili. Thus, when someone buys 'chile powder' they are getting ground fruits of the *Capsicum* plant. But when the consumer gets 'chili powder', it is a mixture of chile powder, cumin, garlic, onions, etc.

We have chosen to use just 'pepper' because of its universal acceptance and because the publisher insisted! Hopefully, in the future as *Capsicum* becomes better known, the word 'pepper' can be retired for the *Capsicum* genus and used only for the *Piper* genus.

INTERNET INFORMATION

The World Wide Web on the Internet provides a new source of information. Many Internet sites have been set up that have valuable information about peppers. Unfortunately, there are also many sites that have wrong or misleading information. In most cases, the information at Internet sites is not reviewed or critically examined, so one must be careful when using the information at the site. An Internet search using some of the most common search engines found more than 300,000 sites that mentioned peppers. For the book we used a few Internet sites to gain valuable information, for example, the FAO site had world production statistics. Where we used information from a site in the book, we referenced the site. The Chile Pepper Institute at New Mexico State University does maintain a site and their address is: www.nmsu.edu/~hotchile/index.html

REFERENCE

Domenici, P. (1983) The correct way to spell chile. *Congressional Record* 129(149), Nov. 3.

Acknowledgements

We gratefully acknowledge the support, information and ideas supplied by the following individuals: Alton Bailey, Louis Biad, Judy, Emily and William Bosland, Emma Jean Cervantes, Margaret Collins, Danise Coon, Dave DeWitt, James Ferguson, James Fisher, Natalie Goldberg, Max Gonzalez, Wendy Hamilton, Kim Harvell, Sharon Hudgins, Jaime Iglesias, Dale Marshall, Don Maynard, Mary K. Riley, Jill Schroeder, Paul G. Smith, Steve Thomas, Randall Tognazzini, Nankui Tong, Peter van Duin, Javier Vargas, Marisa Wall and Yayeh Zewdie.

1

INTRODUCTION

There is a plethora of magnificently coloured and shaped peppers grown worldwide, and the uses for these peppers are as diverse as the fruit types in the *Capsicum* genus. Primarily, pepper fruit is consumed as a fresh vegetable or dehydrated for use as a spice. By volume, red pepper products, pungent and non-pungent, represent one of the most important spice commodities in the world. They add spice flavouring and colour to foods while providing essential vitamins and minerals. In many households, peppers provide the only variety needed to enhance intake of otherwise bland diets. The range of food products that contain pepper or its chemical constituents is broad, and includes ethnic foods, meats, salad dressings, mayonnaise, dairy products, beverages, candies, baked goods, snack foods, breading and batters, salsas and hot sauces. Pepper extracts are also used in pharmaceutical and cosmetic products. In addition to their use as food, condiment and medicine, peppers are also used as ornamentals in the garden.

There is tremendous phenotypic variation in fruit (pod) shapes, sizes and colours, as well as in plant habit among pepper cultivars. A practical way to classify this enormous diversity is to group peppers by species, 'pod-type', cultivar (Bosland *et al.*, 1988). This system of identification is the method most commonly used today. For example, the well-known bell pepper, 'CalWonder', is classified as *Capsicum annuum*, bell pepper, cv. 'CalWonder'. This method enables pepper workers to communicate about peppers in an intelligent manner below the species level.

Peppers were unknown in Europe, Asia and Africa prior to Christopher Columbus landing in the Americas. Columbus may not have been the first person from Europe to reach the western hemisphere, but he is given the credit for introducing pepper to Europe, and subsequently to Africa and to Asia. On his voyage, he encountered a plant whose fruit mimicked the pungency of the black pepper, *Piper nigrum* (L). Columbus erroneously named the *Capsicum* spice 'pepper'. The plant was not the black pepper, but a heretofore unknown plant that was later classified as *Capsicum*. *Capsicum* is not related to the *Piper*

1

(black pepper) genus. In 1493, Peter Martyr (Anghiera, 1493) wrote that Columbus brought home 'pepper more pungent than that from the Caucasus'. The new pepper spread rapidly along the spice routes from Europe into Africa, India, China and Japan. The new spice, unlike most of the plants introduced from the western hemisphere, was incorporated into national cuisines instantaneously (Fig. 1.1).

The eating habits of humans are very conservative. In general, humans are not adventuresome when it comes to foods. We do not stroll through a garden or park and graze on the plants growing within it. We simply do not eat 'unknown' foods. Yet in Europe, Africa and Asia, this new food was integrated into the cuisine without hesitation (Fig. 1.2). This might have been because of the incorrect assumption that the pepper plant was a form of black pepper, *Piper nigrum*. Thus, it was not an 'unknown' food but a known food which until its introduction was only affordable by the nobility. Now peasants could grow 'pepper' and flavour their dishes just like the royal courts.

After Columbus returned to Europe with pepper seed, the extensive trading routes of the Spanish and Portuguese helped spread peppers around the globe. Peppers became the famous Hungarian paprika, which aided Szent-Györgyi in his discovery of ascorbic acid, vitamin C, for which he won the Nobel prize. Birds do not taste or sense the pungency and eat the pods eagerly. Seeds pass through their digestive tracts and thus in Africa were spread further inland, where it became a subspontaneous crop. In West Africa, the native peoples concocted a potion from peppers which ensured eternal youth. Dr Livingstone told Stanley that the native women would sometimes bathe in water with pepper powder added in order to become more attractive. In India and China, peppers began to dominate the cuisine and became the principal spice of the

Fig. 1.1.   Peppers in a Korean marketplace.

region. Pepper quickly became so ingrained in the foods of the region that taxonomists in the 1700s mistook China for the origin of one of the species.

VALUE AND AREA

Peppers are grown in most countries of the world. The production of pepper for spice and vegetable uses has increased by more than 21% since 1994 (FAO, 1997). FAO reports world production of peppers for 1996 to be 14,068,000 metric tons (Table 1.1). Asia is the largest producer (Table 1.1). It is estimated that more than 3 million hectares are grown annually in the world. In most years, China leads world production of peppers with more than 300 hectares harvested annually (Table 1.2). In the United States, New Mexico is the leading state in pungent pepper production, with more than 12,000 hectares under cultivation. California produces the most bell peppers with approximately 10,000 hectares grown annually.

North America and Western Europe are the major importing regions of the world. Peppers are good sources of income to small producers in many developing countries. In 1991, Pakistan, Mexico, India, China and Chile were the principal sources of pepper imports to the United States, while most paprika imports were from Spain, Morocco and Hungary. Exports of pepper from the United States go mainly to Canada, Mexico and Germany.

NUTRITION

Pepper consumption is increasing and may be an important source of vitamins for world populations. The antioxidant vitamins A, C and E are present in high concentrations in various pepper types. Peppers are good sources of many essential nutrients (Table 1.3). Pepper produces high amounts of vitamin C,

Table 1.1. World production of peppers. (Source: FAO, 1997.)

	Area harvested ($\times$ 1000 ha)			Yield (kg ha^{-1})			Production (Mt) ($\times$ 1000)		
	1994	1995	1996	1994	1995	1996	1994	1995	1996
Africa	211	216	221	8,080	8,209	8,310	1,702	1,777	1,839
North America	120	127	126	12,832	12,641	12,516	1,536	1,602	1,575
South America	30	28	29	7,782	8,771	8,564	237	247	248
Asia	712	734	744	10,207	10,483	10,621	7,265	7,692	7,905
Europe	146	149	149	16,227	16,494	16,571	2,366	2,449	2,476
World (total)	1,220	1,255	1,271	10,762*	10,987*	11,064*	13,129	13,792	14,068

*Mean.

Table 1.2. The top pepper producing countries in the world in 1996. (Source: FAO, 1997.)

	Production (Mt)
China	5522
Turkey	1110
Nigeria	970
Mexico	904
Spain	835
USA	597
Indonesia	460
Italy	308
Korea	300
Bulgaria	263
Netherlands	230
Egypt	200
Romania	196
Tunisia	190
Ghana	170
Japan	169

provitamin A, E, P (citrin), B_1 (thiamine), B_2 (riboflavin), and B_3 (niacin). They are richer in vitamins C and A than the usual recommended sources. Considerable research has focused on antioxidants in foods for protection from cancer. Hartwell (1971) lists 14 references where *Capsicum* has been cited as a therapeutic agent for cancer.

One medium green bell pepper (148 g) has 30 calories; 7 g of total carbohydrates, which is 2% of the recommended daily allowance (RDA) for adults; 2 g of dietary fibre, which is 8% of the RDA; 4 g of sugar; and 1 g of protein. It also provides 8% of the RDA of vitamin A, 180% of vitamin C, 2% of calcium and 2% of iron.

As mentioned earlier, pepper fruits vary in size, shape, colour, flavour and pungency. This variation is also reflected in their nutritional composition. The nutritional composition is determined by the species, the cultivar, the growing conditions and fruit maturity. Further changes can occur during postharvest handling and storage. Howard *et al.* (1994) reported on the differences between pod types and cultivars within pod types. They found that provitamin A activity and ascorbic acid content increased with maturity in all cultivars.

It is not only their nutritional quality that makes peppers an important food crop. Peppers stimulate the flow of saliva and gastric juices that serve in digestion. It has been said that pepper raises body temperature, relieves cramps, stimulates digestion, improves the complexion, reverses inebriation, cures a hangover, soothes gout and increases passion!

Table 1.3. Nutritional constituents in 1900 g sample of green bell pepper (GBP) and red New Mexican (RNM) pod types.

| | Water (%) | Energy (kcal) | Protein (g) | Fat (g) | Carbohydrate (g) | Fibre (g) | Ca (mg) | P (mg) | Fe (mg) | Na (mg) | K (mg) | Vitamins | | | | | |
												A (IU)	B_6 (mg)	B_1 (mg)	B_2 (mg)	B_3 (mg)	C (mg)
GBP	93	25	0.9	0.0	5.3	1.2	6.0	22.0	1.33	3.0	195	530	0.16	0.09	0.05	0.55	128
RNM	88	40	2.0	0.2	9.5	1.8	1.8	46.0	1.20	7.0	340	770	0.28	0.09	0.09	0.95	242

Fig. 1.2. Peppers drying on a mat.

Vitamin A

Peppers are an important and rich source of vitamin A. Vitamin A, *per se*, is not found in peppers. Pepper contains the provitamins, alpha-, beta-, gamma-carotene, and cryptoxanthin, which are all transformed in the human liver into vitamin A. The daily vitamin A requirement is met by consumption of 3–4 g (1/2 tbsp) ground red pepper (Lantz, 1943) (Fig. 1.3). After total energy deficiency, vitamin A and protein deficiencies are estimated to be the most common dietary problems in the world (Pitt, 1979). In addition, evidence from epidemiological studies indicates that higher intake of carotene or vitamin A may reduce the risk of cancer (National Academy of Science, 1982; Ziegler *et al.*, 1986). Beta-carotene is the most plentiful form of provitamin A and can be cleaved to form two molecules of retinol, the physiologically active form of vitamin A.

Carrot, *Daucus carota* L., is estimated to be the most important plant source of provitamin A carotenes for many in the world. However, vitamin A is one of three essential nutrients consumed in marginal amounts in the United States, especially by Hispanics (USDHEW, 1968–70; Briggs, 1981). A survey

Fig. 1.3. Ground red pepper powder.

in northern Mexico indicated that peppers are among the 20 foods most frequently consumed by the population of this region and, therefore, constitute an essential part of the Mexican diet (Valencia *et al.*, 1983). This same data also revealed that the per capita dietary intake of peppers was about 40 g per day. Thus, pepper intake in Hispanics, which is overlooked in most studies because peppers are considered to be spices not vegetables, contributes a significant amount of vitamin A to their diet. Mejia *et al.* (1988) examined the provitamin A content of several common Mexican pepper types, and found that the ancho pepper had the highest provitamin A content (111.41 ± 28.2 μg retinol equivalents 100 g^{-1}) and the caribe pepper had the lowest (2.2 ± 0.9).

Vitamin C

Peppers are also among the richest known plant sources of vitamin C (ascorbic acid). Vitamin C was first purified from pepper in 1928 by Hungarian biochemist Albert Szent-Györgyi, who later won a Nobel Prize in physiology and medicine in 1937 for his work with vitamin C. A pepper pod can contain six times as much vitamin C as an orange. Pepper pods from the green to the succulent red stage all contain enough vitamin C to meet or exceed the adult RDA. Fresh fruits may contain up to 340 mg of vitamin C 100 g^{-1} of fruit (Sviribeley and Szent-Györgyi, 1933; Jachimoiwiez, 1941). A 156 g serving, equal to one medium-sized fruit, provides 130% of the RDA in vitamin C. Vitamin C content diminishes by about 30% in canned and cooked pepper, and nearly vanishes from dried pepper (Lantz, 1946).

ETHNOBOTANY

Ethnobotany concerns the plant-lore of people and the way people perceive plants. In most societies, ethnobotany has provided beneficial information about medicinal, spiritual and agricultural uses. The native peoples of the western hemisphere have a great wealth of lore about the *Capsicum* species. Many times it is the shaman, or curer of the tribe, that has the most interesting tales of pepper.

Our association with peppers began 10,000 to 12,000 years ago, when people first inhabited the western hemisphere (MacNeish, 1964). Peppers must have possessed mystical and spiritual powers to our ancestors. The person who took the first taste of pepper fruit was rewarded for this adventurousness in gastronomy with a burning sensation. The amazing part is that these people did not avoid this plant of pain but worshipped it as a gift from the gods. Peppers were held in such high regard by the Aztec, the Maya and the Inca, that they withheld pepper from their diets when fasting to earn favours and to please the gods. Today, sorcerers and shaman of the native peoples prescribe peppers, not so much as a cure, but as a preventive measure against future maladies caused when a person is not in harmony with their surroundings or to protect them against future evil witchcraft. Furthermore, the domestication of pepper was not an isolated event. Because there are five different domesticated species, we can infer that peppers were probably domesticated at least five times independently.

Archaeologists and historians have not placed much value on peppers in their discussions. It has been considered a marginal part of pre-Columbian agriculture, not as a source of protein or carbohydrates to sustain a civilization. They have viewed peppers as a minor crop used only for seasoning. Even though the acreage of peppers was small, they played an important part in the daily lives of the early Americans.

The importance of pepper as a seasoning in pre-Columbian times is confirmed by the writings of the Spanish chroniclers of the 16th century. Fray Bernardino de Sahagún described Aztec food in 1569. Sahagún wrote that in the market there were 'hot green peppers, smoked peppers, water peppers, tree peppers, flea peppers, and sharp-pointed red peppers'. Sahagún described in detail the pungency and aromas of peppers in the marketplace. The Aztecs classified peppers into six categories based not only on level of pungency (high to low), but also on the type of pungency (sharp to broad). Thus, pepper was more than a simple spice. Unlike black pepper which adds only bite, pepper accentuates the flavour of the food dish. To further illustrate the importance of the flavour differences among the different pepper types, Sahagún described how each pepper was used in dishes. He wrote 'frog with green pepper, newt with yellow pepper, tadpoles with small peppers', and so on.

Pepper was one of the most common tribute items in pre-Columbian times and continued to be so after the Spanish conquest. Tribute was a form of taxation by the Incas and Aztecs, and was adopted by the Spanish after their arrival in Mexico. Pepper tributes to dignitaries were common. In 1550, Don Juan de Guzmán, the governor of Coyoacan in the Valley of Mexico, received 700 peppers per week as tribute (Durán, 1588). Peppers were used as currency in Central America until the 1900s. McBryde (1933) reported that pepper was used as 'small change' in exchanges. In 1945, he stated that 12 peppers were worth about 4 or 5 onions or a 'pinch' of salt (14 g) (McBryde, 1945).

In his book, *Gathering the Desert*, Gary Nabhan (1985) describes a portion of the creation myth from the Cora Indians of Mexico's west coast as written by a Franciscan priest in 1673. In the narrative, God makes a man (Narama) and a woman (Uxuu). Narama is the patron saint of salt, mescal and chile. God made a fiesta and Narama is among the last to arrive. He comes naked and covered with salt. After everyone is seated, Narama presents himself at the table and takes salt from his face and sprinkles it upon the food. Then he reaches down, and his testes turn into chile pods. He begins to sprinkle this spice on to all of the foods. This crude action annoys all the other patrons, who angrily scold Narama. To this he replies that if the others could provide the fruits, fish, fowl, seeds and vegetables that are the basic staples of the fiesta, why could not he provide something that these foods needed to be truly tasteful? He declares that there is nothing so necessary as salt and chile. The guests try the chile with the food and become enthusiastic. From that day on, they knew in their hearts that chile would always be in their diet.

In 1609, Garcilaso de la Vega wrote about peppers in Inca society in his *Royal Commentaries of the Incas*. He wrote that Incas worshipped pepper as one of the four brothers of their creation myth. Pepper was the brother of the first Inca king. The pepper pods were perceived to symbolize the teachings of the early ancestors. Within their society, peppers were holy plants and had to be avoided when fasting. Peppers were used as money and even today purchases can be made in Peru with a handful of peppers. Incas decorated pottery and clothing with peppers. An obelisk from the Chavín culture in Peru has a carving of the black caiman, a mythical creature, and in its claws are the pods of peppers. This suggests that Incas believed there were inherent spiritual powers in the peppers.

In Peru, the Inca ruler Huayna Capac ordered the nobles of Cuzco to go and purchase coca and peppers so that he could perform the fiesta of Purucaya as a tribute to his mother. The Purucaya was a very special fiesta in Inca life. This ceremony was performed to show great respect to the deceased. The ceremony would be comparable to canonizing someone today (Betanzos, 1576).

With the availability of other fruits and vegetables high in nutritional value, why did our ancestors domesticate and grow peppers? It may have been the myriad fruit shapes and colours of peppers that interested them (DeWitt and Bosland, 1996). It may also have been the flavour and aroma which are as important in peppers as colour and form. The most potent volatile in peppers is

2-methoxy-3-isobutyl-pyrazine, the 'bell pepper' smell. Humans can detect this compound at the lowest concentration of any compound known to human-kind. However, in spite of these attributes of peppers, it was probably pungency that caused our ancestors to invest considerable time and energy in growing and improving peppers.

The popularity and spiritual essence of peppers is associated with pungency. Some have argued that pungency should be one of the five main taste senses, along with bitter, sweet, sour and salty. Pungency is produced by the capsaicinoids, alkaloid compounds that are unique to the genus *Capsicum*. It is pungency that allowed pepper to be used for more than just food in pre-Columbian times. Pepper was probably first used as a medicinal plant. The Mayans used peppers to treat asthma, coughs and sore throats. In Columbia, the Tukano group use peppers to relieve a hangover. After a night of dancing and drinking alcoholic beverages the Tukanos pour a mixture of crushed pepper and water into their noses to relieve the effects of the festivities. The Aztecs and the Mayas mixed pepper with maize flour to produce *chillatolli*, a cure for the common cold. The Teenek (Huastec) Indians of Mexico use pepper to cure infected wounds (Alcorn, 1984). The pepper fruit is rubbed into the wound and can produce pain so severe that the patient passes out. The Teenek believe that the pepper kills the 'brujo' (evil spirit) causing the illness. Some other uses include putting red crushed fruits on feet to cure athletes foot fungus, and to cure snakebite by making a drink from boiled green fruits.

Medicine men use it in a maceration mixed with aguardiente to give as a purge for dogs to make them good hunting dogs. The Jivaro apply the fruit directly to a toothache. In Piura, a fruit infusion is considered antipyretic, tonic and vasoregulatory. The decoction is used as a gargle for sore throat or pharyngitis, and the tincture is applied to insect bites, mange, haemorrhoids, and rheumatism. The Rio Apaporisa natives take the fruits for flatulence, and use small quantities of powdered fruit when breathing is difficult. It is also used for scorpion stings, toothache, haemorrhoids, fever and flu (Duke and Vasquez, 1994). The Aztecs placed a drop or two of pepper juice on a toothache to stop the pain. In 1590, while living in Mexico, the priest Acosta remarked 'when the pepper is taken moderately it helps and comforts the stomach for digestion, but if they take too much it has bad effects'. He also stated that pepper was bad for the 'health of young folks, chiefly their souls, because consuming pepper provokes lust'. In the Amazon the indigenous people believe that to become a good blowgun shooter, one must chew and eat slowly half a dozen fruit before breakfast for 8 days.

Pepper is an important part of the daily lives of the Barama River Caribs in British Guyana. Peppers are used as vegetables in many of their dishes. They are cooked with meat and vegetables in the pepper pot, and are also eaten raw as garnish in sandwiches of meat or fish and cassava bread. As a remedy for headache, a feather is rubbed in the pulp of a pepper and drawn across the eyes or pepper juice is dropped in the eyes. Shamans drink a mixture of pepper juice and water to induce the psychic state necessary for communication with the supernatural powers (Gillin, 1936).

It has been reported that the leaves and stems of the pepper plant have been effective as an antibiotic, carminative, rubefacient, stimulant, stomachic and a vesicant. The capsaicinoids are proclaimed to cure or be effective as an antidote to poison arrows, aphrodisiac, asthma, baldness, boils, breathlessness, bronchitis, cancer (nose), carcinoma, carminative, chest-cold, chills, cholera, ciguatera, CNS stimulant, collyrium, conjunctivitis, cough, counterirritant, dandruff, diaphoretic, diarrhoea, digestive, dropsy, dysmenorrhoea, dyspepsia, earache, epithelioma, evil eye, fumitory, gargle, gonorrhoea, headache, impotency, intoxication, jaundice, lumbago, malaria, migraine, neuralgia, pharyngitis, phthisis, psychedelic, rheumatism, rubefacient, scrofula, shark repellant, skin stimulant, snake bite, skin sores, sore throat, stimulant, stomachic, styptic, toothache, tuberculosis, tumour, typhus and vertigo (Beckstrom-Sternberg *et al.*, 1994).

Pepper was not only used as a reward or to cure sickness, but it was also used as punishment. In the Mendocino Codex (a codex was a method of recording information usually in the form of paintings) the daily life of Aztecs was described. The Codex described a common form of punishment for children. It has a drawing of a father punishing his 11-year-old son by making the boy inhale smoke emanating from dry peppers roasting on the hearth. In the same drawing, a mother threatens her 6-year-old daughter with the same punishment. Today, we have 'pepper spray' which is standard issue for many police departments in the United States to control unruly criminals.

MEDICINAL USE

Since pre-Colombian times pepper has been used as a medicinal plant. Today, peppers are one of the most widely used of all natural remedies. A survey of the Mayan pharmacopoeia revealed that tissues of *Capsicum* species (*Solanaceae*) are included in a number of herbal remedies for a variety of ailments of probable microbial origin (Cichewicz and Thorpe, 1996.) As stated earlier, it may have been pepper's medicinal use that caused the indigenous peoples of the Americas to domesticate peppers (Fig. 1.4).

Medicinally, capsaicin is being used to alleviate pain. At present, it is the most recommended topical medication for arthritis. At nerve endings a neurotransmitter called substance P informs the brain that something painful is occurring. Capsaicin causes an increase in the amount of substance P released. Eventually, the substance P is depleted and further releases from the nerve endings are reduced. A decrease in substance P also helps to reduce long-term inflammation. Inflammation can cause cartilage break down. Cream containing capsaicin is used to reduce the pain associated with post-operative pain for mastectomy patients and for amputees suffering from phantom limb pain.

Fig. 1.4. Indigenous woman by her prized pepper plant.

Prolonged use of the cream has also been found to help reduce the itching in dialysis patients, the pain from shingles (*Herpes zoster*), and cluster headaches (Carmichael, 1991.) The repeated use of the cream apparently counters the production of substance P in the joint, hence less pain.

Other medical uses of peppers will be found. Currently, about ten research papers a month are published on the medicinal use of peppers (Bosland, 1997). No doubt this trend will continue as new applications are learned (Fig. 1.5).

CONCLUSION

In the 21st century, peppers will play an important role in food use, medicine and many other areas. Consumption is increasing, new uses are being discovered and the interest of the general public for this crop seems insatiable. The future will bring new peppers to the consumer. Just as the coloured bell peppers, the non-pungent jalapeño and new ornamentals have increased awareness of peppers, new flavours and new pepper types will bring forth a cornucopia of novel products for the consumer.

There is a need for a text that consolidates so many aspects of pepper research, production and uses. Each chapter of this book provides up-to-date

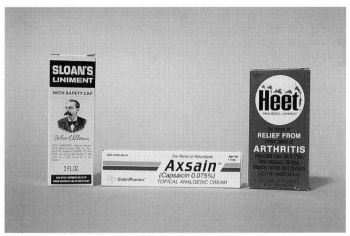

Fig. 1.5. Medicines using pepper pungency (capsaicin) as an active ingredient.

information on specific topics of pepper. The book is not intended to be the complete tome on *Capsicum*, but a versatile book that is informative and authoritative without necessarily being exhaustive. We hope that our approach with the chapters has produced a book that is useful to the horticulturist, the grower, the researcher and the layperson who has an interest in *Capsicum*.

2

TAXONOMY, POD TYPES AND GENETIC RESOURCES

INTRODUCTION

Capsicum species are members of the *Solanaceae*, a large tropical family that includes tomato, potato, tobacco and petunia. They are not related to *Piper nigrum*, the source of black pepper, nor are they related to the Guinea pepper or grains of paradise, *Aframomum melegueta*.

All *Capsicum* species originated in the western hemisphere, except for *Capsicum anomalum*. *C. anomalum* is a monotypic species that originated in Asia. Taxonomically, it is distinct from the other *Capsicum* species and it is questionable whether it belongs in the genus.

NUMBER OF SPECIES

There has been considerable debate on the number of species in the genus. The first literary references on the classification of pepper are from 16th century botanical books. Before Linnaeus wrote his seminal work *Species Plantarum* in 1753, several authors tried to classify peppers. Morrison, in 1699, published *Plantarum Historiae Universalis Oxoniensis* and had 33 variants for pepper. In 1700 Tourneforte gave the genus its name, *Capsicum*, and listed 27 species. Linnaeus reduced *Capsicum* to two species, *annuum* and *frutescens*. Later in 1767, he added two more species, *baccatum* and *grossum*. Willdenow (1798) described the species *pendulum* followed by Ruiz and Pavon (1790) describing the species *Capsicum pubescens*. In the first half of the 19th century attempts were made to clarify questions related to the taxonomy of the genus. Dunal (1852) described 50 species in the genus and listed another 11 as possible species. By the end of the 19th century, the names of more than 90 species were listed within the genus. Irish (1898) recognized only two species, *annuum* and *frutescens*, as in initial work by Linnaeus. The two species concept was widely accepted until 1923 when L.H. Bailey made the argument that because *Capsicum annuum* was a perennial in the tropics or could be grown in a greenhouse as a perennial, there was only one species, *frutescens*. In addition, confusion reigned because some authors agreed to the one species concept but used *annuum* instead of *frutescens*. Heiser and Smith (1953) recategorized the genus into four species, *annuum*, *frutescens*, *baccatum* and *pubescens*, and in 1957 they determined that *Capsicum chinense* was a unique species (Smith and Heiser, 1957) and brought about the current list of five domesticated species.

BIOLOGICAL/MORPHOLOGICAL SPECIES

Capsicum is endowed with a multitude of fruit forms, colours and sizes. Without genetic knowledge, the early taxonomists named *Capsicum* species based mostly on this fruit morphology. Scientists define species in two ways, the biological species and the morphological species. Each is based on a different set of criteria for establishing the species. A taxonomist, when establishing a morphological species, examines floral traits and looks for similarities and differences in the flower structure. Seeing none, a taxonomist would group the two populations of the plant together under one species name. A biological species is defined as a population or series of populations within which free gene flow occurs under natural conditions. This means that the two populations of plants must be able to produce fertile and healthy progeny in subsequent generations. If there is free genetic exchange between two populations, they would be considered to be the same species. For the most part, within *Capsicum*, the morphological species concept has been used to establish species. As biological information on the *Capsicum* species is obtained, the species within *Capsicum* may be changed dramatically.

Saccarod and La Gioria (1982) found that when a *C. annuum* accession from Colombia was crossed to *C. annuum* accessions from Mexico and New Mexico, USA, abnormal chromosome pairing occurred. Reduced fertility in the progeny was caused by several translocations that had occurred between the Colombian population and the other two populations. Geographic isolation of populations has allowed *C. annuum* species to begin a differentiation process that could possibly lead to two different species.

Today, the genus *Capsicum* consists of at least 25 wild species and five domesticated species (Table 2.1), but this is only an estimate. It is anticipated that new species could be discovered and named in the future. The most recent taxonomy of *Capsicum* above the species level is:

Kingdom	*Plantae*
Division	*Magnoliophyta*
Class	*Magnoliopsida*
Order	*Solanales*
Family	*Solanaceae*
Genus	*Capsicum*

Table 2.1. The described species of *Capsicum*. Source: IBPGR (1983).

Tubocapsicum:	*C. anomalum*	
Pseudoacnistus:	*C. brevifolium*	
Capsicum:	*C. annuum*	*C. baccatum*
	var. *aviculare*	var. *baccatum*
	var. *annuum*	var. *pendulum*
	C. buforum	*C. campylopodium*
	C. cardenasii	*C. chacoense*
	C. chinense	*C. ciliatum*
	C. coccineum	*C. cornutum*
	C. dimorphum	*C. dusenii*
	C. eximium	*C. frutescens*
	var. *tomentosum*	*C. galapagoense*
	C. geminifolium	*C. hookerianum*
	C. lanceolatum	*C. leptopodum*
	C. minutiflorum	*C. mirabile*
	C. parvifolium	*C. praetermissum*
	C. pubescens	*C. schottianum*
	C. scolnikianum	*C. tovarii*
	var. *flexuosum*	
	C. villosum	

The eminent Argentinean taxonomist, Armando Hunziker (1979), has classified the genus into three sections: *Tubocapsicum* with one species *C. anomalum*; *Pseudoacnistus* with one species *Capsicum brevifolium*, and *Capsicum* with the remaining species.

The genus name, *Capsicum*, is most likely from the Latin, *capsa*, meaning 'satchel'. However if could be from the Greek, *kapto*, meaning 'to bite'. Pungency is one of the most characteristic traits of the *Capsicum* genus. Even though bell peppers are non-pungent, they still belong to *C. annuum*. A single mutation causes the loss of ability to produce the capsaicinoids, the pungent compounds. Most *C. annuum* cultivars are pungent. However, there are several wild species, e.g. *Capsicum ciliatum* (see Fig. 2.1) and *Capsicum lanceolatum*, which never have pungent fruit. Some taxonomists have suggested that the non-pungent fruit species from the wild should not be included in *Capsicum*. Heiser and Smith (1958) suggested that the genus should be limited to those species with a pungent, shiny, non-pulpy berry. Those plants with soft pulp-filled, non-pungent berries should be excluded from the genus. Recent studies are indicating that the chromosome number for non-pungent species is $n = 13$, compared with $n = 12$ for the pungent species (Tong and Bosland, 1997). If this cytological and biochemical evidence holds true for all non-pungent species, it may provide the catalyst to move the non-pungent '*Capsicum*' species to a new genus.

Fig. 2.1.   *Capsicum ciliatum,* a wild relative of pepper.

WILD PEPPER FRUITS

When the terms wild, cultivated and domesticated are used with peppers they actually represent a continuum of human–plant relationships. At one end of this continuum are wild plants, which grow outside the human-disturbed habitat and cannot successfully invade human-disturbed areas. This seems to be true of *C. lanceolatum*; a wild species that has only been found in virgin rainforest of Guatemala. Further along the continuum are the semi-domesticated cultivars such as *Capsicum annuum* var. *avirculare*, the wild chiltepin (Fig. 2.2). Next are the domesticated plants which have evolved into new forms under continued human manipulation so that they may have lost the ability to reproduce themselves without human care, e.g. bell peppers.

All wild peppers share similar fruit traits and a common characteristic of being associated with birds. The fruits are small, erect, with a soft pedicel trait that allows red ripe fruit to be pulled easily from the calyx. This permits easy removal by frugivorous birds, and eventual dissemination of the seeds. The red colour is attractive to birds and it seems birds cannot taste or feel the capsaicinoids. The capsaicinoids are secondary metabolites that must have

Fig. 2.2.   Wild chiltepin, the mother of all peppers.

evolved to ward off pests (mammals) and encourage dispersal by birds. The small pods of the wild species are commonly called 'bird peppers' in languages all over the world because of this association.

DOMESTICATED SPECIES

From native species, early humans in Mexico, Central America and South America domesticated at least five species independently in different regions. Even so, some wild species are still harvested in the wild and utilized by humans. *C. pubescens, C. baccatum, C. annuum, C. chinense* and *C. frutescens* are considered to be the five domesticated species (IBPGR, 1983). An understanding of each of the domesticated species will illustrate the evolution and possible origins of each species. Scientific studies indicate that the domesticated species belong to three distinct and separate genetic lineages. Although the barriers between the gene pools may be breached through human hybridization, this rarely happens in nature.

Several of the species in the genus can be grouped into species-complexes. The complexes contain species that allow for genetic exchange, albeit with some difficulty, between the species. Each complex is made up of the domesticated species and their wild relatives. Each can be viewed as a primary gene pool of genetic diversity. The pubescens-complex consists of *C. pubescens*, *Capsicum eximium* and *Capsicum cardenasii*, while the baccatum-complex has *C. baccatum, Capsicum praetermissum* and *Capsicum tovarii*. The annuum-complex consists of *C. annuum, C. frutescens, C. chinense, Capsicum chacoense* (see Fig. 2.3) and *Capsicum galapagoense*.

C. pubescens Complex

The pubescens-complex consists of relatively unknown peppers. *C. pubescens* was originally described by Ruiz and Pavon (1790) from plants cultivated in Peru. *C. pubescens* probably originated in the highlands of Bolivia and, according to Heiser (1976), it was domesticated *c.* 6000 BC, making it one of the oldest domesticated plants in the Americas. *C. pubescens* was domesticated in the Andes, where it is often called locoto or rocoto. It is found from Mexico to Peru, growing in the Andean South America and the Central American highlands in small family plots. It is grown on a very limited acreage in the rest of the world.

Instead of white flowers, *C. pubescens* has purple flowers with large nectaries. The presence of conspicuous leaf pubescence and black seeds readily distinguishes this pepper from any of the other species. *C. pubescens* is a large, shrubby, herbaceous plant that can grow to 12 m and has lived up to 10 years in the tropical Americas. This pepper is adapted to cooler temperatures, 4.5–15.5°C. However, a myth has been perpetuated that the species is frost tolerant. This is not true; however, when the plant is established (> 1 year old) it will reshoot after a frost. It is most likely that stored carbohydrates in the roots enable the plant to regrow. The fruit types of *C. pubescens* vary in shape and colour, but do not have the same tremendous diversity of pod types as *C. annuum*. The fruits may be elongate to spherical, and may or may not have

Fig. 2.3. *Capsicum chacoense,* a wild relative of pepper.

a pronounced neck (see Fig. 2.4). The range in mature fruit colours includes red, orange and yellow. The fruit has thick flesh, and does not store or dehydrate very well. Some accessions are self-compatible, while others are self-incompatible.

   Taxonomically, *C. pubescens* belongs to a taxa including *C. eximium* and *C. cardenasii*. *C. eximium* has a botanical variety *tomentosum*. *C. eximium* var. *tomentosum* has more leaf pubescence than the standard species. *C. cardenasii* is a self-incompatible species, while *C. pubescens* has populations that are self-compatible and self-incompatible. *C. eximium* is self-compatible.

C. baccatum Complex

Capsicum baccatum represents another distinct complex. *C. baccatum*, known in the vernacular as *ají*, extended its range from southern Brazil west to the Pacific Ocean, and became a domesticated pepper of choice in Bolivia, Ecuador, Peru and Chile. In South America, *C. baccatum* is the most commonly grown species. *C. baccatum* has cream coloured flowers with yellow, brown or dark green spots on the corolla. One type of *C. baccatum* called 'puca-uchu' grows on vinelike plants in home gardens. Two botanical varieties of *C. baccatum* are recognized: *C. baccatum* var. *baccatum* and *C. baccatum* var. *pendulum*. *C. baccatum* var. *baccatum* is the wild form with *C. baccatum* var. *pendulum* containing the domesticated forms. In older literature, *C. baccatum* var. *microcarpum* is listed, however, it is now considered to be the same as *C. baccatum* var. *baccatum*. As many different pod types (in relation to shape, colour and size) exist in

Fig. 2.4. *Capsicum pubescens* fruits.

C. baccatum as in *C. annuum* (see Fig. 2.5). Fruits vary in pungency from very mild to fiery hot. They embody unique aromatics and flavours. *C. baccatum* is the pepper of choice when making ceviche (marinated fish).

On the basis of crossing between *C. baccatum* var. *pendulum* and *C. baccatum* var. *baccatum*, it was determined that they were one species. The crosses produced fertile offspring in the F_1 and F_2 and had normal chromosome pairing. Karyotypic analysis revealed that the chromosomes of the two species were identical. Some crosses did produce a reduction in fertility indicating that genetic isolation was beginning to be developed (Eshbaugh, 1963). *C. baccatum* has at least two wild relatives, *C. praetermissium* and *C. tovarii* (Tong, 1998). Sometimes *C. praetermissium* is listed as a variety of *C. baccatum*; however, it is a distinct species.

C. annuum Complex

The three other domesticated species, *C. chinense, C. frutescens* and *C. annuum*, share a mutual ancestral gene pool and belong to the annuum species-complex. Some authors have suggested that at the primitive (wild) level it is impossible to differentiate among the three species. Those taxonomists who use the morphological species concept have grouped all three together as one species. However, biological studies seem to present convincing evidence that they are truly three separate species. Even with studies on sexual compatibility and chromosome behaviour between *C. chinense* and *C. frutescens*, the legitimacy of the two species is still questioned. Cytogenetic studies show aberrant chromosome pairing between the interspecific crosses which supports

Fig. 2.5. Diversity of *Capsicum baccatum* fruits.

the differentiation into three species. A single chromosome translocation differentiates *C. chinense* from *C. annuum* (Egawa and Tanaka, 1986). Each species was domesticated independently: *C. annuum* in Mexico, *C. chinense* in Amazonia (or possibly Peru) and *C. frutescens* in southern Central America. Today, these three species are the most commercially important peppers in the world.

C. annuum pod types are usually classified by fruit characteristics, i.e. pungency, colour, shape, flavour, size and use (Smith *et al.*, 1987; Bosland, 1992) (see Fig. 2.6). The most likely ancestor of *C. annuum* is the wild chiltepin (*C. annuum* var. *aviculare*). It has a wide distribution, from South America to southern Arizona, but the cultivated *C. annuum* was first domesticated and grown in Mexico and Central America. By the time the Spanish arrived in Mexico, the Aztecs had already developed dozens of varieties. Undoubtedly, these peppers were the precursors to the large number of varieties found in Mexico today. The pod types within *C. annuum* are discussed later.

The *C. chinense* species, like all *Capsicum* species, originated in the western hemisphere. However, the Dutch physician, Kikolaus von Jacquinomist, who named this species in 1776 said that he named it after its 'homeland', China (Jacquin, 1776). It is still a mystery why he thought China was its place of origin. Because of taxonomic convention that the first name given to a species is used, the misnomer *chinense* is still attached to this western hemisphere native. This is the species most often grown in Brazil, where the species diversity is enormous and underexploited (Cheng, 1989). *C. chinense* is popular in all tropical regions and is the most common species in the Caribbean. The diversity

Fig. 2.6. Diversity of *Capsicum annuum* fruits.

in fruit shape in this species can equal that found in *C. annuum* (see Fig. 2.7). Fruit can be extremely pungent and aromatic, with persistent pungency when eaten. The habanero has the distinction of being the hottest pepper in the world. The plant sets two to six fruits per axil.

The *C. frutescens* (synonyms: *Capsicum minimum* and *Capsicum fastigiatium*) species has fewer cultivars than *C. chinense* and *C. annuum*. In Brazil, there is a cultivar named 'melagueta'. This pepper is not related to *Aframomum melegueta*, the melegueta or Guinea pepper, which is related to ginger. There are fewer names for the undomesticated varieties of *C. frutescens* than for the undomesticated varieties of other species. No large fruited *C. frutescens* has ever been found in an archaeological site in Central or South America, but ethnobotanists speculate that the domestication site was probably Panama, and from there it spread to Mexico and the Caribbean. Some varieties of *C. frutescens* are grown in Africa, India and the Far East, where they are still called bird peppers. Many cultivars listed as *C. frutescens* in the trade are often actually *C. annuum* cultivars.

POD TYPES

Classification of peppers, as with any multifarious group of cultivars, is confusing. Within *C. annuum*, *C. chinense* and *C. baccatum* there is a great number of pod types. These pod types are distinguished mostly by their characteristic pod shape, but can also be differentiated by their use, fruit colour, pungency level, aroma and/or flavour. These subspecific categories are used by the pepper industry to aid in supplying the correct pepper for the proper product.

Fig. 2.7. Diversity of *Capsicum chinense* fruits.

A horticulturist and a taxonomist differ in how they classify plants. The taxonomist classifies down to a species, or at most, to a variety level. The taxonomist generally does not recognize differences below the variety level. However, horticulturists are interested in identifying crops below the species level. Within a species, horticulturists classify crops into races, cultivars, and specifically with peppers, into pod types. From a taxonomic point of view there are no significant differences between the cultivars of *C. annuum* to warrant species differentiation. However, from an industry, plant breeder's or horticulturist's point of view there are qualities, such as flavour, which are not taxonomic characters at all, but have great significance to the industry. Classification below the species level is a horticultural approach, not a taxonomic one.

Classification

The early species, *Capsicum grossum*, was based on fruit shape (Linnaeus, 1737). Sturtevant (1919) proposed classifying peppers based on the type of calyx and described seven classes. Various schemes and names for the classification of peppers below the species level have since been proposed. An early classification system had botanical varieties within *C. annuum* based on fruit shape. Examples of this system were *C. annuum* var. *cerasiforme* (cherry), *C. annuum* var. *grossum* (bell) and *C. annuum* var. *longum* (New Mexican). Today, this system is no longer recognized or used. The current system is to use genus, species, variety, pod type and cultivar (Bosland *et al.*, 1988). At least 50–75 distinct pod types are known. Some of the most recognizable pod types are bell, jalapeño, cayenne, New Mexican, yellow wax, ancho and mirasol.

The best way to understand pepper pod types is to use an analogy of breeds of cattle. There are various species of the cow, such as *Bos taurus* and *Bos gaurus*, in the same way as there are various species of *Capsicum*. Within *B. taurus*, there are breeds of cows, such as Hereford – raised for meat and Holstein – raised for milk producing capabilities. This is similar to ancho, bell and New Mexican pod types in *Capsicum*. Within the Hereford breed there is a polled and a horned variety. This is analogous to cultivars of New Mexican, 'New Mexico 6-4' and 'NuMex Big Jim'. For example, a horticulturist would label a New Mexican cultivar, this way: *Capsicum annuum* var. *annuum*, New Mexican, cv. 'NuMex Big Jim'. Because cultivars can cross-pollinate, thousands of different peppers exist around the world. Just as one can have cross-breeds and mongrels in animals, peppers can have types that do not fit established pod type categories. Many have only a common name, making identification difficult. In Mexico, there are more than 200 common names for the approximately 15 pod types cultivated commercially (Laborde and Pozo, 1982).

C. annuum Pod Types

A major division between the different *C. annuum* pod types (see Fig. 2.8) is to classify the fruits as pungent or not. The non-pungent (sometimes called 'sweet') pod types include bell, pimiento, Cuban and squash. There are pepper types that have both pungent and non-pungent cultivars, such as yellow wax and cherry. The pungent *C. annuum* varieties include cayenne, New Mexican, jalapeño, serrano, ancho, pasilla, mirasol, de Arbol and piquin. However, there are also cultivars of jalapeños and New Mexican that are non-pungent.

Bell
The bell group may be the most economically important pod type. It also has the largest number of cultivars. It is interesting to note that Wafer the Pirate in 1699 referred to bell peppers in his diary about his adventures on the Panama Isthmus (Wafer, 1699). It is highly unlikely that the bell pepper he refers to is the same one being grown today. In North America bell peppers refer to fruits that are blocky and about 10 cm long and wide. A square shape with a flat bottom is preferred. 'California Wonder' is one of the oldest cultivars and is typical of the pod type. In Europe, an elongated bell pepper, La Muyo, is grown. The most distinguishing characteristics of the La Muyo are the non-flat blossom end and the two–three celled fruits, instead of the four-celled fruit of the North American type. Bell pepper cultivars can begin as green, purple, yellow, or white and ripen to shades of red, orange, yellow, green or brown. There are also some cultivars that are pungent, e.g. 'Mexibell'.

Pimiento
Pimiento, sometimes spelled pimento, is characterized by a heart-shaped, thick-walled fruit that is green when immature and red at maturity. The fruit are non-pungent and the wall flesh is sweeter tasting than bell peppers. Pimiento is used in processed foods, such as pimiento cheese and stuffed olives, but can be eaten fresh. Allspice, *Pimenta dioica*, usually known as pimento or Jamaican pepper outside the United States, is not related to *Capsicum* (Bosland, 1992).

There are two types of pimientos grown in the USA, the oblate fruit of the tomato type and the conical-shaped fruit. Tomato type cultivars include 'Sunnybrook' and 'Early Sweet Pimiento', while 'Perfection' and 'Sweet Meat Glory' are cultivars of the conical-shaped fruit type. 'Perfection Pimiento', one of the most recognizable pimiento cultivars, was first listed in 1914 in Riegel's seed catalogue. The catalogue cited it as a sport of a cultivar received from Valencia, Spain.

Squash/tomato/cheese
Pepper fruits of the squash, tomato or cheese types differ from bells and pimiento in that the fruits are generally flat. Fruit diameter ranges from 5 to

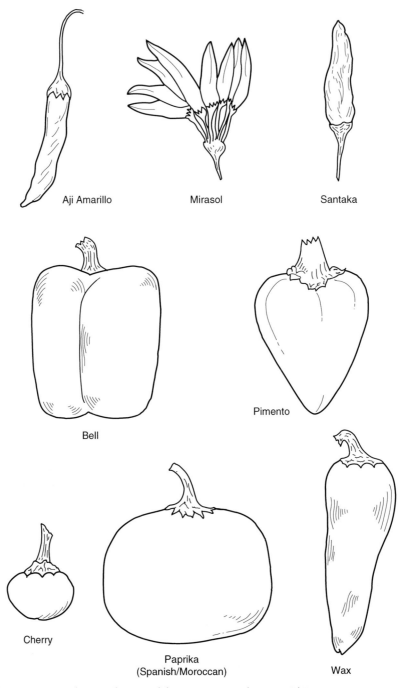

Fig. 2.8.　Fruit shapes of some of the common pod types within peppers.

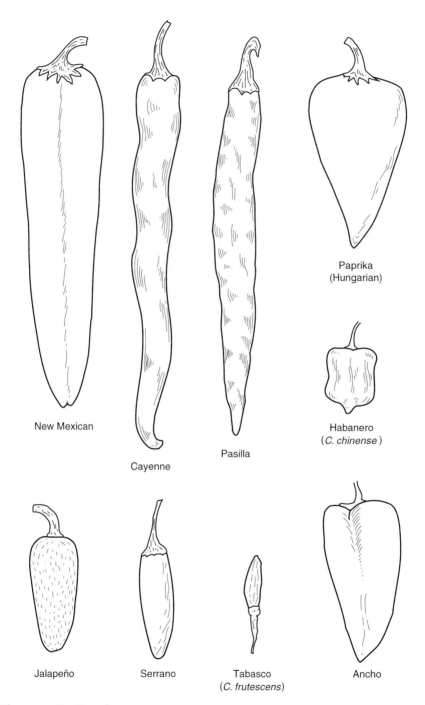

New Mexican

Cayenne

Pasilla

Paprika
(Hungarian)

Habanero
(*C. chinense*)

Jalapeño

Serrano

Tabasco
(*C. frutescens*)

Ancho

Fig. 2.8. *Continued.*

10 cm wide. The fruits begin as green and turn a red, yellow or orange at maturity. They are non-pungent like bells and pimiento fruits. They are usually pickled, but may also be eaten fresh in salads. In Guatemala, they are called 'chamborate' and are used extensively in November and December to enhance the typical foods, frambre, and Christmas tamale.

Yellow wax
Wax fruits are yellow when immature, with a waxy appearance, and turn orange, orange-red or red at maturity. In Spanish, they are referred to as 'güero', because 'güero' means light-skinned or blond. There are two main forms within this group, a long-fruited type and a short-fruited type. The long-fruited types are known as 'Hungarian Wax' or 'Banana Peppers'. The fruit length is about 10 cm and width 4 cm. Common cultivars in the long-fruited form are 'Sweet Banana', 'Feherozon', 'Giant Szegedi' and 'Corbaci'. The short-fruited types are known as the cascabella or just as yellow wax. The fruits are seldom longer than 5 cm in length and are less than 2 cm wide. Some common cultivars are 'Floral Gem', 'CalOro', 'Cascabella' and 'Fresno'. The cultivars may be pungent or not. Wax fruits are often pickled but can be used fresh in salads or relishes.

Cherry
Cherry types have small, round or slightly flattened, immature green fruit that turn red at maturity. As the name suggests, the shape of the pod is similar to that of a cherry. Depending on the cultivar grown the fruit is either pungent or not. Culinary use is the same as for the wax pepper. An ornamental potted plant called 'Jerusalem cherry' is not a pepper, but *Solanum pseudocapsicum*. It is important to identify 'Jerusalem cherry' correctly because its fruits are poisonous.

Paprika
Paprika is a peculiar pepper category. It is not a pod type in international spice trade or in the United States, but it is a product. Any non-pungent, dried, red powder is paprika to the international spice trade. This mild powder can be made from any type of *C. annuum* that is non-pungent and has a brilliant red colour. However, in Europe, there are pepper pod types which are paprikas. This is because in the Hungarian language 'paprika' means *Capsicum* or pepper and has been referred to as 'red gold' by Hungarian farmers (Somos, 1984). Paprika may be pungent in Hungary, but it is always non-pungent in international trade. Even though paprika is considered to be a spice product in international trade, it is consumed as an important vegetable in European and North African diets. In Europe, paprika is made from two principal fruit types: (i) a round fruit about the size of a peach and called Spanish or Moroccan paprika; and (ii) a longer, more conical and pointed type grown in the Balkan countries, called Hungarian paprika. The basic types in Spain are the 'dulce'

(sweet, no pungency), the 'ocal' or 'agirdulce' (bittersweet) which have a slight pungency, and 'picante' (hot) which has a noticeable pungency. Because paprika can be made from different pod types, Spanish, Hungarian and US paprika each taste slightly different.

Most of Hungary's paprika is produced in the regions around Kalocsa and Szeged. The spice paprika is grown exclusively outdoors, while the vegetable paprika is grown outdoors in the summer and in greenhouses in the cooler months. Hungary produces a wide range of paprikas from very mild to very hot. The Hungarian type paprika is also produced in Macedonia and Bulgaria. The fruit shape of Hungarian paprikas varies from tomato-shaped to triangular or heart-shaped to thin, elongated cones.

Chiltepin/chile piquin

These are the pepper pods found growing naturally in the wild. They are the 'mother of all chiles'. There are many names applied to this pod group and it is used for the small pods in other species. A common name is 'bird pepper' because of the fondness birds show towards it. The fruits are small, 2 cm long and 1 cm wide. The term 'tepin' is used to differentiate the round fruit shape, while 'piquin' is oval or bullet-shaped. The green fruit is pickled, while the red form is dried and used as a seasoning.

Ancho, mulato and pasilla

There is much confusion with the names for the fruits in the ancho group. Ancho fruit is pungent, heart-shaped, pointed, thin-walled and has an indented stem attachment. The immature fruit colour is dark green. If fruit colour is red at maturity, ancho types retain the name ancho, but if it is dark brown at maturity, it is called mulato. Poblano is used by the United States' produce industry for any green ancho fruit. Technically, poblano is a specific ancho grown in Puebla, Mexico (Laborde and Pozo, 1982). Ancho is not widely grown outside Mexico, but is the pepper of choice for making chile rellenos (stuffed peppers) in Mexico.

The ancho pod type was developed in pre-Columbian times, and today has retained many of the pod characteristics of the earlier cultivars. There is variability in plant height, leaf size, form and colour among the various ancho cultivars. The typical ancho pod is 8–15 cm long, conical or truncated in shape and cylindrical or flat, with a marked depression at the base. The apex is pointed and slightly flat. The skin is a dark, deep green, but will turn red when fully mature. It has 2–4 lobes. Mulato is very similar to ancho but matures to a dark chocolate brown instead of red.

Adding to the confusion of names for these peppers, the mulato is mistakenly called pasilla in California. The reason for this ambiguity is that pasilla means raisin in Spanish, so any dried, wrinkled chile could be a pasilla. The true pasilla is a long, slender, dried pepper pod. The pasilla fruit is cylindrical and undulating. Pods are 15–30 cm long and 2.5–5 cm wide. They are dark green

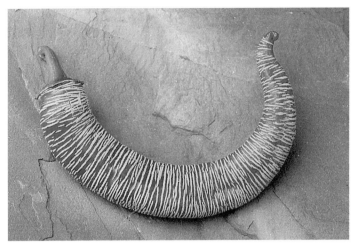

Fig. 2.9.   Cayenne fruit exhibiting extreme corkiness.

and turn brown at maturity. The green fruit of pasilla is called 'chilaca' in the Mexican marketplace and is used like green pepper. The word 'chilaca' also now refers to the green fruit of the New Mexican type. The mature pasilla is dehydrated, then used in mole sauces.

Cayenne

Cayenne (see Fig. 2.9) named either for the city or the river in French Guiana, has red mature fruit and is characteristically wrinkled. Pod length is 13–25 cm by 1.2–2.5 cm wide. The pod may be crescent or irregular in shape. It is highly pungent (30,000–50,000 Scoville units[1]). It is grown commercially in Africa, India, Mexico, Japan and the United States. In the United States it is grown in Louisiana, New Mexico and Texas where it is made into a mash with salt to be used to make hot sauces. It can be dried and ground into a powder, commonly known as 'red pepper' (Bosland, 1992).

Chihuacle

Chihuacle is a rare pod type grown only in southern Mexico. The name suggests a pre-Colombian domestication. Chihuacles vary in shape, but usually

1   Pepper pungency is expressed in Scoville Heat Units (Scoville, 1912). The Scoville Organoleptic Test was the first reliable measurement of the pungency of peppers. This test used a panel of five human subjects, who tasted a pepper sample and then recorded the heat level. A sample was diluted until pungency could no longer be detected. The organoleptic method or taste test has been the standard method for pungency analysis. Although this method is widely used it has limitations. Tasters must be trained and their ability to test many samples is restricted by the heat of the test solutions. Taster fatigue is a real phenomenon and tasters are also not able to distinguish between the different capsaicinoids. Therefore, the Scoville Organoleptic Test has been replaced with instrumental methods.

measure 5–7.5 cm in length and 4–6 cm in width. The fruits are thin-walled and range in appearance from that of a miniature bell pepper to pods that are broad shouldered tapering to a point. Immature fruit are green, ripening to yellow, red or even a black colour, hence the names chihuacle amarillo, chihuacle rojo and chihuacle negro. The differently coloured chihuacles are used to produce the unique mole sauces for which the Mexican state of Oaxaca is famous.

Cuban and pepperoncini

The fruit of the Cuban type has large, irregular, thin walls. 'Cubanelle' and 'Aconcagua' are two common cultivars within the Cuban group. With the pepperoncini, there are two types, Italian pepperoncini and Greek pepperoncini. The Italian pepperoncini has pods that are dark green in colour. The Greek pepperoncini is also called Tuscan. The Greek pepperoncini are really a medium-light green colour when picked fresh. When the peppers are brined they are treated with sodium bisulphite to remove the green colour. After desalting, the peppers are prepared in a solution of turmeric and/or FDA Food Color yellow #5 with vinegar for consumer packaging. Pepperoncini and Cuban types have very little pungency or none at all.

Costeño

Costeño peppers are grown commercially in Guerrero, Mexico. There is a lot of variability in plant type because the Costeño pepper is a landrace. The plant can grow to 1.5 m high, and has many branches starting at the ground level. The pods are long and either conical or oval in shape. They vary in size from 2 to 15 cm long by 1 to 3 cm wide. The body of the pod is cylindrical and very wrinkled, with some having very deep constrictions in the skin. The outer skin is thin and brittle and when it dries it becomes transparent. The pods are predominantly light green in colour, almost yellowish; when they are fully mature they turn light red. They are very pungent.

Mirasol

This group consists of peppers called mirasol, guajillo and cascabel. The fruits are translucent and thin-walled. This group is used in the dry form. The mirasol type has erect fruit, hence the appropriate name 'mirasol', which means looking at the sun. Some of the new mirasol cultivars do have pendulate fruit. Pods are 7–10 cm long, and 1–2 cm wide and slightly curved. The guajillo type is a rich burgundy-red colour when dried. It has the shape of a small New Mexican type. The pods are 11 cm long and 2.5 cm wide. The guajillo is also known as 'pulla'.

Cascabel

This pod type is sometimes listed under the mirasol group. The cascabel is similar in shape to the cherry, but has a thinner wall. The fruit is most often used in the dry form and it is the dry form that gives the pod its name. The dried fruits have the distinct characteristic of the seeds 'rattling' in the pod, hence its name, cascabel (rattle). The pods are spherical and shiny, when dry they are mahogany to brown in colour. The cascabel should not be confused with the cultivar 'cascabella' which is in the wax group.

de Arbol

The name, 'de Arbol', is derived from the resemblance of the pepper plant to a tree, although it grows only 60–152 cm high. The fruits are 5–8 cm long, 0.5–1 cm wide, and translucent when dried. The calyx end of the fruit is narrow and tapered, which distinguishes it from mirasol. The fruits are larger than the chile piquin. Mexican common names for this chile are pico de pajaro (bird's beak) and cola de rata (rat's tail).

Jalapeño

The jalapeño pepper was named for the town of Jalapa, Mexico where it was originally marketed. However, it was not originally grown there, but was imported from the surrounding regions. Fruits are thick-walled, conical-shaped, dark green when immature (turning red at maturity) and highly pungent. However, jalapeños of other colours have been developed in the United States. The cultivar 'NuMex Piñata' matures from light green to yellow to orange and finally to red as it ripens (Votava and Bosland, 1998). They are principally used as a spice and condiment. Most of the jalapeño crop is preserved by canning or pickling, while a small amount is dehydrated in either the green or red stage. Fruit skin may show a netting pattern called corkiness. Corkiness on the fruit is a desirable trait in Mexico but is undesirable in the United States. The thick fruit walls keep the pod from drying naturally. Mature red jalapeños are dried by smoking them over mesquite or a hardwood, and the product is called 'chipotle', which is not a separate criteria.

Serrano

This pod type probably originated in the mountains of northern Puebla and Hildago, Mexico, hence its name 'Serrano', literally from the highland or mountain. It has cylindrical fruits 5–10 cm long, 1 cm wide, with medium-thick walls and no corkiness. The immature fruit colour ranges from light to dark green. Fruits are red, brown, orange or yellow when mature. The pungency level is higher than jalapeño. Serrano is the pepper of choice for making Pico de Gallo, a salsa-type relish.

New Mexican

The New Mexican pod type is also called long green chile or 'Anaheim'. In fact, the pod type is New Mexican, and 'NuMex Joe E. Parker' and 'Anaheim' are cultivars within this pod type. The New Mexican pod type was developed in 1894 when Fabian Garcia at New Mexico State University began improving the local peppers grown by the Hispanic gardeners around Las Cruces, New Mexico. Green and red pepper represent two developmental states of the same fruit. The long green pods that turn red are the pepper of choice for Mexican-style cooked sauces in the United States. 'Anaheim' seed originated in New Mexico and was brought to Anaheim, California, where it was widely cultivated. Numerous landraces have evolved in New Mexico and are named for the growing area, such as Chimayo, Dixon and Velarde.

New Mexican green pepper is roasted and peeled for fresh consumption, and for canning or freezing. The skins must be removed before using. If pods are left on the plant to be harvested at the red stage, they are usually dried and ground into pepper powder (paprika if non-pungent). All New Mexican type peppers grown today gained their genetic base from cultivars first developed at New Mexico State University.

Santaka/hontaka

This pod type is typical of the pungent peppers from Japan. The fruits are 7 cm long and ¾ cm wide, and set in clusters on the plant similar to mirasol types. The fruits are very hot, but the pungency dissipates rapidly. The fruits are used in the red ripe stage, where they are dried and used as a seasoning.

Ornamental

Ornamental peppers as potted plants are popular in Europe and are gaining in popularity in the United States (Armitage and Hamilton, 1987; Bosland *et al.*, 1994). Ornamental peppers are not really a 'pod type', but a unique class of peppers. Covered with bright red fruits during the holiday season, ornamentals are often called Christmas peppers (see Fig. 2.10). Although edible, ornamentals are grown primarily for their unusual pod shapes or for their dense foliage and colourful fruits. Merits of ornamental pepper as a potted plant include easy seed propagation, relatively short cropping time, heat and drought tolerance, and excellent keeping quality. Most ornamental peppers are pungent, but not poisonous. The subtle flavours associated with other pod types are missing in most ornamentals.

C. chinense Pod Types

There are a myriad of pod types within the *C. chinense*. The Amazon basin has the largest amount of diversity in *C. chinense* pod types. *C. chinense* is also the most commonly grown species in the Caribbean. The two most familiar pod types within *C. chinense* are the habanero and Scotch bonnet, which differ in pod shape. These two names are sometimes wrongly used interchangeably. For example, if the native language is Spanish, habanero is preferred, and when

Fig. 2.10.   An ornamental pepper.

English is the native language, Scotch bonnet is commonly used. Research and discussion on the various pod types within *C. chinense* is limited. There are many pod types that lack any description. There are also names such as 'country pepper' that apply to more than one pod type. A brief description of some of the more common pod types follows.

Habanero
The habanero pod type is described as lantern-shaped, orange or red at maturity, and is very hot. The habanero chile was originally grown on the Yucatan Peninsula of Mexico and in Belize. Commercial production has expanded to Costa Rica and the United States. The pods can mature into orange, yellow, white or red. The pod is 6 cm long and 2.5 cm wide at the shoulders. One cultivar, 'Red Savina', which has red mature fruit colour, claims to be the world's hottest pepper at 577,000 Scoville units. The fruits are used fresh in salsas, cooked directly in dishes or fermented to make a hot sauce.

Scotch bonnet
The Scotch bonnet is shaped like a tam-o-shanter, thus the name 'bonnet'. It is grown extensively in Jamaica. It usually matures to yellow, white, red or orange. A chocolate colour is rare. It is as pungent as the habanero and is the same size. It is used in similar ways.

Datil
The Datil pod type is grown in the area of St Augustine, Florida. It is thought that the datil pepper was introduced to the area by Minorcan settlers in the 1700s. However, a more plausible explanation is that it was introduced by

trade with the Caribbean islands. The pods are green, maturing to orange or yellow. They are 9 cm long and 2 cm wide at the shoulders.

Charapita/pimento de cherio
The pod type 'Charapita' or 'Aji Charapa' is found in the Peruvian jungle, close to the city of Iquitos, where the people are called 'charapas'. The pod is very small (0.6 cm diameter) and spherical. 'Aji Charapa' is thin fleshed, very pungent, and matures into red or yellow colours. A similar small pod type from Brazil is called 'pimento de cherio'. The word 'cherio' in Portuguese translates to odourous. A feature of all *C. chinense* fruits is the strong aroma which has been described as apricot-like to some, while others have claimed the smell is like lanolin! The fruit of the 'pimento de cherio' and the 'charapita' resemble chiltepins, but are usually orange or yellow at maturity instead of red. The pods are upright, round and about 5–8 mm in diameter.

Additional pod types
There are several other *C. chinense* pod types. In Brazil there is the 'Cheira Bell'. It is 2 cm long and 1 cm wide. The fruit is green with a purpling from anthocyanin, and matures to red. The 'Cumari o passarinho' from Brazil is 3 cm long, and 1 cm wide. The pod matures from green to orange. In Africa, one of the few *C. chinense* pod types named is the 'fatalli'. The pods are 8 cm long and 4 cm wide. The pods are pointed at the tip and change from green to yellow at maturity.

In the Caribbean, the 'Congo' is grown on Trinidad and Tobago. It is the largest *C. chinense* pod type. The pods have thick walls and mature from green to red. The pods are 6 cm long and 7–8 cm wide. On the island of Puerto Rico and in the West Indies, the most common pod type is the 'Rocotillo'. This pod is similar in shape to the Scotch bonnet, but has a very long pedicel. The heat of this pod type is less than that of the standard Scotch bonnet.

In Panama, a named pod type is the 'Aji Chombo'. Even though most 'aji' peppers are *C. baccatum*, this pod type is truly a *C. chinense*. It is 6 cm long and 2.5 cm wide.

In Peru a very popular *C. chinense* pod type is the 'Ají Panca' (CPI, 1997). It is the second most common pepper variety in Peru, after the *C. baccatum* cultivar 'Ají Amarillo'. 'Ají Panca' is grown mainly near the coast. The fruit measures 8–13 cm long and 2.5–3 cm across. 'Ají Panca' has the same fruit shape and appearance as the 'Ají Amarillo', but is a deep red to burgundy colour when ripe. The 'Ají Limo' has small fruits that measure 4–8 cm long by 2.5–3 cm wide. It ripens into a deep red, yellow or orange. When dried, this pepper becomes tapered and wrinkled. It is mostly grown and used on the northern coast of Peru. The 'Aji Pucomucho' is a wild variety found in the Peruvian jungle. The fruits are small, thin, elongated and pointed, and mature into a bright-yellow colour.

C. frutescens Pod Types

C. frutescens does not have as many named cultivars as *C. annuum* or *C. chinense*. In Africa and Asia, most peppers that are called 'bird pepper' are *C. frutescens*. One of the most common cultivars of *C. frutescens* is 'Tabasco'. The 'Tabasco' fruits are 2.5–5 cm long by 0.5 cm wide, yellow or yellow-green (turning red at maturity) and highly pungent. The red fruit is the ingredient in Tabasco sauce®.

'Malagueta' is another prevalent cultivar, especially in Brazil. In Africa, it is called the 'Zimbabwe Bird'. It is not related to the true melegueta pepper (*Aframomum melegueta*) which grows in Africa.

C. baccatum Pod Types

The Spanish word 'ají' commonly refers to fruits of *Capsicum baccatum* in South America. The Spanish imported the phonetic, a¢ hee, from the native Arawak peoples of the Caribbean to Peru. Also in Peru, the descendants of the Incas still call them by their Quechuan word, 'uchu'.

Aji Amarillo
The 'Aji Amarillo' is the most common *C. baccatum* pepper in Peru (CPI, 1997). In the United States, it is sometimes called 'Yellow Pepper' or 'Yellow Peruvian Pepper', or 'Escabeche'. The pods are 10–15 cm long and a deep orange colour when mature. The thin-fleshed pods have a fruity flavour with berry overtones and a searing, clear pungency. This pepper is the first choice when making ceviche. This pod type has been known in Peru since ancient Inca times, where it is represented in drawings and pottery.

The 'Aji Ayucllo' is a wild pepper variety found in the Peruvian jungle, near the Chanchamayo and Villa Rica Valleys. The fruits are small, thick-fleshed and oval-shaped with a moderate pungency. The 'Ají Norteno' or the 'northern pepper' is popular in the northern coastal valleys of Peru. The fruits mature to yellow, red or orange. They are 8–10 cm long and 2 cm wide. The fruit shape is slightly curved and tapering to a point. The Virú and Lambayeque valleys, about 1000 km north of Lima, are the main production areas.

C. pubescens Pod Types

This species has numerous landraces grown from the Andes of Peru to the highlands of Mexico. It is grown extensively in courtyards and kitchen gardens. The pods combine the suavity and juiciness of the bell pepper and the heat of a habanero. The different fruit shapes have not been described in detail. However, two main pod types are 'manzano' (apple-shaped, red-coloured) and 'peron' (pear-shaped, yellow-coloured). Other names for peppers in this species are 'siete caldos' (hot enough to season seven soups), caballo (the heat kicks like a horse), locoto and rocoto (from the Quechuan language).

Other Pod Types

The descriptions above do not define all the pod types in the world. Many of the known groups lack a complete and thorough description to allow for even limited pod type classification. The most glaring omissions in the literature are the Asian and African pod types. Even though peppers were introduced into African and Asian countries after Columbus, there has been extensive selection for fruit shape, flavour, and use. India, China, Japan, Korea and Thailand grow extensive hectarage of peppers, but information on the specific pepper types grown is not available.

One may also expect novel pod types in the future. Just as plant breeders in the past developed new pod types, plant breeders in the future will develop new pod types. As plant breeders develop peppers, their selection will undoubtly create changes that will be so significant that a new name will need to be applied to the peppers to describe them properly. Thus, 'pod types' is a dynamic classification system and as a better understanding of pod types is gained, more pod types will be named.

ADDITIONAL SPECIES

The other approximately 20 *Capsicum* species lack extensive study on their biology. Many of the known wild species have restricted distribution. These species may contain genes for adaptation to unusual environmental conditions as well as disease resistance. A complete evaluation of the *Capsicum* genepool would require that all the wild species be represented in collections so that their potential for plant breeding can be established. This may be impossible because the natural habitat for these species is in danger of being lost. Collection or information may never be accomplished before their extinction. Tropical deforestation is among the most massive and urgent environmental problems facing *Capsicum* germplasm resources.

Wild species of *Capsicum* will be useful in breeding for disease resistance and, in addition, will be used to increase the nutritional quality, yield and adaptability to stress environments. The enhancement of commercial cultivars by exotic germplasm is dependent on the availability of living material. The disappearance of *Capsicum* is directly linked to the disappearance of the rainforest where *Capsicum* originated. The genetic diversity of *Capsicum* can be saved only through the use of several strategies. One approach is to locate areas that may still harbour *Capsicum* species and protect the areas from further development.

The preservation of *Capsicum* genetic resources in natural sites of occurrence must be encouraged. When possible it is desirable to set up *Capsicum* genetic resource reserves in conjunction with relevant biosphere resources and other protected areas, such as the chiltepin preserve near Tucson, Arizona. Another approach to preserving the genetic resources of *Capsicum* is to enlarge and conserve germplasm in base and active gene banks. The USDA Plant Introduction Station at Griffin, Georgia, is a good example. The improvement of appropriate storage facilities for germplasm and the financial support of those operations is very urgent. It is also imperative to aid the active collections of Latin America, where *Capsicum* is native.

USDA CAPSICUM GERMPLASM COLLECTION

The United States National Plant Germplasm System houses an extensive *Capsicum* germplasm collection at the Southern Plant Introduction Station located in Griffin, Georgia. This collection contains approximately 3000 *Capsicum* accessions including accessions from all over the world. Passport data are recorded upon arrival of the seed at the facility and a USDA Plant Introduction Number (PI#) is assigned. Evaluation data are subsequently entered in the Germplasm Resources Information Network (GRIN) after they have been received. The collection is the source of germplasm for breeding and research programmes throughout the world. Many have been evaluated for descriptors.

GRIN is a centralized computer database system that manages plant genetic resources. It is through GRIN that scientists can locate plants with specific characteristics and then obtain them for research purposes. The database is designed to permit flexibility to users in retrieving information. There are three ways to gain access to GRIN data. One is to write for a hard copy of the data, another way is to use PC diskettes. But a more efficient and expedient way is with a computer. Using the Internet, one can reach the GRIN database. The database contains pertinent information about a particular accession, from its native habitat to the most recent characteristic and evaluation results. The database allows scientists access to information on a more extensive collection of germplasm samples. It also reduces the possibility of overlooking a potentially valuable sample.

The web site is available at http://www.ars-grin.gov. This address is for the home page of the National Plant Germplasm System. Once there, choose the 'plants' option. There it is possible to search the *Capsicum* database.

ADDITIONAL GERMPLASM COLLECTIONS

Globally, there are several other pepper collections. Many of the accessions in each collection are duplicates of material kept in other collections. The most active collections are the Asian Vegetable Research and Development Centre (AVRDC) Tainan, Taiwan; Centro Agronomico Tropical de Investigations y Ensenanza (CATIE), Turrialba, Costa Rica; Centre for Genetic Resources (CGN), Wageningen, The Netherlands; and the Central Institute for Genetics and Germplasm, Gatersleben, Germany.

3

BOTANY

INTRODUCTION

The physiology of peppers is discussed in two chapters. This chapter discusses the morphology, growth and development of vegetative structures of the pepper plant (see Fig. 3.1), while in Chapter 4, aspects of seed formation, production, germination, dormancy, etc., will be addressed.

In their native habitats, peppers are grown as tender perennials. In many parts of the world, however, they are grown as annuals. Their morphology is similar to that of tomato, which are members of the same botanical family. There are significant differences between the two crops. Pepper roots are fibrous, and top growth of shiny, glabrous, simple leaves is generally more compact and more erect than in tomato. Cultivars may vary from the 'normal' description, so intraspecific variation as well as interspecific variation must be taken into account. There are mutants for most of the morphological floral features. There are even mutants that are non-flowering (see Fig. 3.2). Listing and describing known pepper mutants is beyond the scope of this book, therefore the reader is referred to Daskalov and Poulos (1994).

EMBRYO AND SEEDLINGS

Pepper is a dicotyledonous plant of epigaeic germination. The cotyledons may differ in shape and size, but a typical cotyledon is wide in the middle and gradually narrows towards the apex and the base part (see Fig. 3.3). Some variant seedlings have three cotyledons instead of the normal two. This trait appears to be genetically controlled.

The main root and young rudiments of the later branch roots can be differentiated in the seedling. The main root axis consists of a vigorously developed main root with lateral roots located on the axis in an evenly funiform distribution. The roots of pepper have a deep taproot, unless the root tip is damaged (see Fig. 3.4). The root tip is damaged when seedlings are grown in transplant containers. The root tip can also be damaged when the plants are direct-seeded and the taproot hits a hard pan layer in the field. When the taproot is damaged, new lateral roots develop from the primary taproot system. The lateral roots develop from the main root in two opposite rows. The root system of a fully developed pepper plant resembles a dense 'tassel'.

The majority of the roots are located near the soil surface. Hortizontally they spread to a length of 30–50 cm and grow 30–60 cm in depth. In modern cultivars, the root mass is relatively small in comparison with the rest of the plant. As a general rule, root weight is approximately 10% of the total plant weight. Juvenile plants have a higher ratio of root weight to top growth. This ratio gradually declines as the plant increases its foliage and stem percentage. Adventitious roots are rare in peppers.

STEM AND LEAVES

The shoot system is highly variable. The young stems are angular, becoming circular in cross-section as they mature. *C. baccatum* has a distinctive squarish stem. The stem may have anthocyanin along its length and anthocyanin may or may not be present at the nodes. The stem can be glabrous, pubescent or a gradation between the two extremes. There are indeterminate types that grow like vines and semi-indeterminate types where the plant grows and, as fruit sets, the plant slows its growth. A true determinate type, as found in processing tomatoes, does not exist in pepper. However, the fasciculated plant habit has branches that end in a fruit cluster. Examples of this branching habit include the mirasol and the santaka pod types (see Fig. 3.5).

Most *C. annuum* cultivars develop a single stem with 8–15 leaves before the appearance of the first flower. The number of leaves that appear before the first flower seems to be controlled by temperature and cultivar genotype (Deli and Tiessen, 1969). With the development of the first flower bud the plant branches at the apex into two or more shoots. Each shoot bears one or two leaves, terminates in a flower and then divides into two second-order branches. The two lateral branches form a dichasium and the terminal bud is transformed into a floral apex (Shah and Patel, 1970). One of the dichotomic branches is sometimes suppressed, especially in the third and higher branches, so that the branch system tends towards a sympodium (Shah and Patel, 1970).

The leaves of pepper have variation in size, shape and colour. Most are simple, entire and symmetrical. They can be flat and smooth or wrinkled and glabrous or subglabrous. Some are pubescent as in the serrano types, or the species *C. pubescens* (see Fig. 3.6). The leaf blade may be ovate, elliptic or lanceolate. The leaves are usually green, but types with purple, variegated or yellowish colour are known. The leaf petiole can be short or long depending on species and cultivar. Leaves develop either in clusters, singularly in a spiral

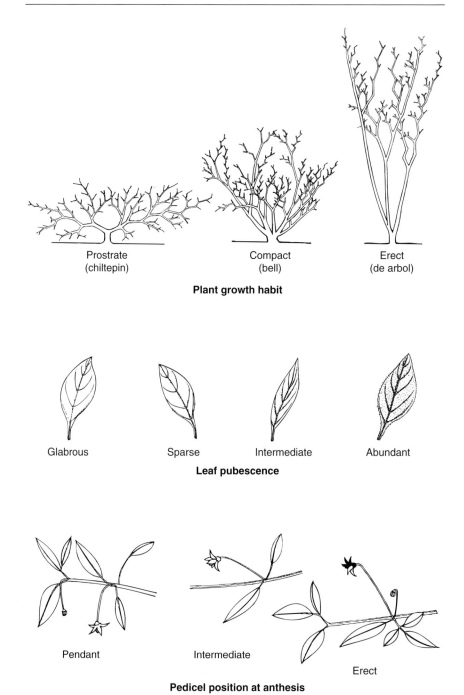

Prostrate
(chiltepin)

Compact
(bell)

Erect
(de arbol)

Plant growth habit

Glabrous   Sparse   Intermediate   Abundant

Leaf pubescence

Pendant   Intermediate

Erect

Pedicel position at anthesis

Fig. 3.1. *Capsicum* descriptors after the International Board for Plant Genetic Resources. (IBPGR, 1983.)

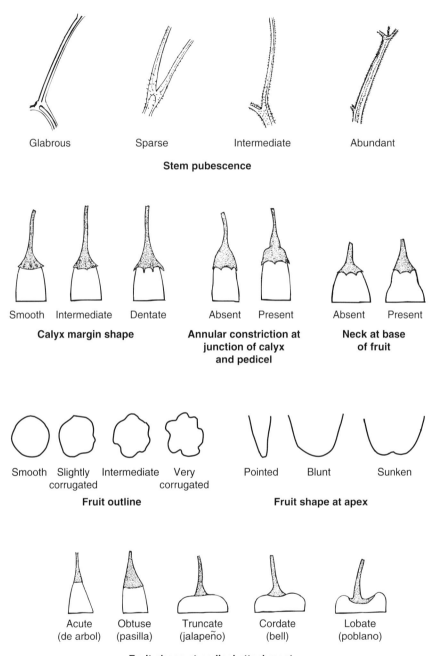

Glabrous   Sparse   Intermediate   Abundant

Stem pubescence

Smooth   Intermediate   Dentate

Calyx margin shape

Absent   Present

Annular constriction at junction of calyx and pedicel

Absent   Present

Neck at base of fruit

Smooth   Slightly   Intermediate   Very
corrugated   corrugated

Fruit outline

Pointed   Blunt   Sunken

Fruit shape at apex

Acute   Obtuse   Truncate   Cordate   Lobate
(de arbol)   (pasilla)   (jalapeño)   (bell)   (poblano)

Fruit shape at pedicel attachment

Fig. 3.1. *Continued.*

Fig. 3.2.   Pepper mutants.

Fig. 3.3.   Pepper seedlings.

system, or by pairs in opposite position. On the main axis, leaves are, as a rule, of spiral arrangement. Leaf apex is acuminate, but can vary to acute or obtuse. The leaf base either gradually narrows into the petiole or is abruptly acute. The number of stomata per leaf ranges from 120–190 mm^{-2} for leaves grown in full sun to 35–70 mm^{-2} for leaves grown in the shade (Schoch, 1972).

DEVELOPMENT

The juvenile period in pepper is short and is expressed only when the apical development is retarded by environmental factors, chemical sprays or pinching. Depressing apical development stimulates the sprouting of lateral buds on

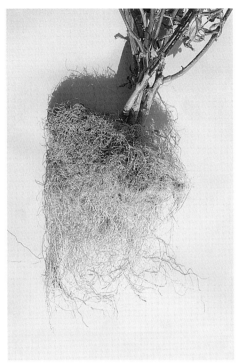

Fig. 3.4. Pepper root system from a container, illustrating the lack of a taproot and the profusion of fibrous roots.

Fig. 3.5. Fruit of 'NuMex Mirasol', illustrating the cluster (fasciculated) and upright fruit habit.

Fig. 3.6.   Leaves of *Capsicum pubescens*, illustrating the hairiness of the plant and reason for name.

the main stem below the first branching node and causes the plant to 'bush-out'. Buds originating from different nodes on the main axis differ in their readiness to flower (Rylski and Halevy, 1972). The higher buds, that is those closer to the first flower primordium, will flower earlier. This is true even if the flowers have already been formed there. The flowering of lower buds will be delayed. Lateral shoots that develop from the most juvenile buds produce four or five leaves before their first flower. The number of leaves produced gradually decreases towards the apex, and the shoots developing from buds close to the apex produce only one or two leaves before flowering. This character was retained when detached single node explants were grown in culture (Rylski and Halevy, 1972).

FLOWER DIFFERENTIATION

Pepper flower differentiation does not appear to be affected by day length. The literature on the subject of day length effect is limited. Most accessions will flower with a day length of 10 h or more. The most important factor determining flower differentiation is air temperature, especially night temperature. The flower opens within the first 3 h after sunrise and is open for less than 1 day (see Fig. 3.7). The anthers may open from 1 to 10 h after the flower opens, but frequently they fail entirely to dehisce.

Nectar is produced and accumulates in the nectary at the base of the ovary. The quantity of nectar depends on many factors but the genotype of the cultivar seems to be the most important. The flowers are visited by bees for both the

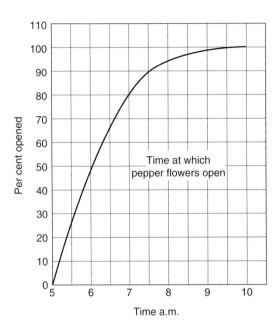

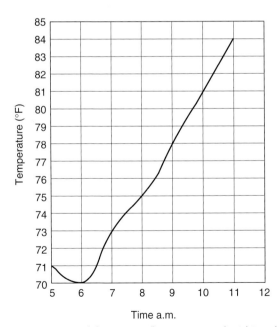

Fig. 3.7. The major portion of the pepper flowers opened within 3 h after sunrise.

Fig. 3.8. A bee pollinating a pepper plant in the field.

nectar and the pollen (see Fig. 3.8). Older cultivars and wild species have conspicuous nectar drops. The nectar drops are in the middle of the lower section of the flower petals associated with a pore that is visible with a hand-lens. Many new cultivars do not produce significant amounts of nectar, thus bee visitation can be low. Bee visitation is also dependent on the relative attractiveness of competing flowering plants in the locale. It has been reported that flowers of *Capsicum* have no odour. However, the flowers of *C. pubescens* can fill a greenhouse with a sweet floral scent when in full bloom.

Most *Capsicum* species have flowers that are self-compatible. *C. cardenasii* is an exception among the *Capsicum* species in being self-incompatible, as well as some accessions of *C. pubescens*. Erwin, in 1937, measured the effect of pollination on fruit set in *C. annuum*. He found that only 46% of self-pollinated flowers set compared to 71% that were left to open pollinate by bee activity. When flowers were left to self, Nagarathnam and Rajamani (1963) obtained only a 6–11% fruit set of the flowers present. The degree of cross-pollination is not as rare as textbooks lead one to believe (Odland and Porter, 1941; Tanksley, 1984a). In many locales, cross-pollination is predominant. The amount of cross-pollination depends on several factors but can range from 2 to 90% (Pickersgill, 1997).

FLOWER

The typical *Capsicum* flower is pentamerous, hermaphroditic and hypogynous. The corolla is rotate in most species with 5–7 petals which are 10–20 mm long. The notable exceptions are *C. cardenasii* and *C. tovarii*, which have a campanulate corolla. The diameter of a *C. annuum* flower is 10–15 mm across. The wild species have smaller flowers. The corolla is either a solid colour or has spots. The corolla spotting is a useful trait in species delineation. The flower colour is dependent on the species, but most *Capsicum* species have whitish flowers. In *C. annuum*, there are a small number of accessions that have purple corollas. *C. frutescens* has a greenish flower. There are a number of species that have purple corollas as the primary colour. These are known as the purple flowered species: *C. eximium*, *C. pubescens*, and *C. cardenasii*.

Flowers are usually solitary at the axils of the branches for *C. annuum*, but other species have multiple flowers at the nodes, e.g. *C. chinense*. Some accessions, for example the mirasol, have clusters of flowers at the node (Bosland and Gonzalez, 1994). The cluster type is associated with the fasciculated gene, which causes multiple flowers/fruits to form at a node.

C. annuum (see Fig. 3.9) start flowering with a single flower at the first branching node; there can be exceptions where two flowers can be found at some nodes. Then a flower forms at each additional node, a geometric progression. Generally, more than 100 flowers develop on one plant. The rate of fruit set is negatively correlated to the number of fruits developing on the plants. When the plant has set several fruits, the rate of flower production decreases. Fruits from early flowers are usually larger and have greater red colour and pungency content at maturity. Fruits do not set when mean temperatures are below 16°C or above 32°C. However, flowers drop when night temperatures are above 24°C. Maximum flower set occurs when the day and night temperature is between 16°C and 21°C. Fruit set may be stalled if temperatures rise above 32°C after several flowers have set and fruits are developing. This causes a split in the fruit setting continuum, and is called a split-set. Early yield is determined by the first flowers setting fruits. A delay in fruit set can reduce yields and may cause fruit to set high on the plant, which makes plants more prone to wind damage and lodging as they mature. Fruit normally reaches the mature green stage 35–50 days after the flower is pollinated.

Flower Drop

Flower drop in pepper has been reported as a common problem during hot weather. Erickson and Markhart (1997) reported that temperature is the primary factor in the decrease of fruit production. Decreased production is due to flower abortion and not to decreased flower initiation or plant growth. When night air temperatures are 32–38°C, fruit does not set. However, when the

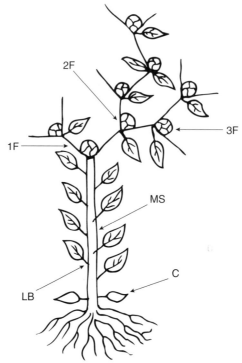

Fig. 3.9. Schematic diagram of *Capsicum annuum* plant development: C, cotyledons; MS, main stem; 1F, first terminal flower; 2F and 3F, second and third node flower; LB, nodes.

night air temperature is lower than 16–21°C there is a marked increase in fruit setting. Cochran's (1932) reason for the poor fruit set at high temperatures was excessive transpiration by the plant. Dorland and Went (1947) thought it was insufficient sugar translocation.

Pepper flowers have five stamens with anthers varying in colour from bluish to yellow to white depending on the species, and alternate in position with corolla lobes. There is an equal number of stamens and corolla lobes. The anthers are not united as in the tomato or potato. The anthers are 1.2–2 mm wide and 2–4 mm long. They dehisce laterally along a line that runs the whole length of the anther. A single stigma which may vary from slightly shorter than the anthers to much longer is present in peppers. The filaments may be white or violet depending on the species and are 1.8–3.5 mm long. The relative position of the stigma and the anthers may vary considerably among cultivars. The range in positions varies from short-styled with the stigma below the level of the top of the anthers to long-styled with the stigma extending above the plane of the top of the anthers. The frequency with which the stigma protrudes above the anthers is greatest in the 'wild' type peppers, such as the piquin types, and least among the most 'domesticated' bell peppers or large fruited types.

The pistil comprises an ovary with a longitudinal diameter of 2–5 mm and a transverse diameter of 1.5–5 mm containing 2–4 carpels, a style 3.5–6.5 mm long, and a capitate and lobed papillate stigma that has a mean diameter slightly greater than that of the style. The period of receptivity of the stigma is 5–7 days, depending on environmental conditions (Cochran and Dempsey, 1966). There is no perfect synchronization in the pollen release and the receptivity of the stigma. Flowers are protogynous, that is at the bud stage (flower has not opened) the pollen is not mature but the stigma is receptive. Pepper breeders take advantage of this condition to make controlled crosses.

The ovary is variable in shape, but the shape of the fruit is determined by the growth of the ovary after anthesis (Kano *et al.*, 1957). In blocky and conical fruits, the ovary is round or slightly elliptical, in oblong fruits, prismatic or conical. On the apex of the ovary there is a minute conic protuberance holding the style. The style is a fragile thin or thick filiform formation that ends in the stigma. The stigma is funnel shaped, segmented and has 'globules' at its edge. On the surface, it is finely papillate. The style varies in length by variety. Because the pepper style varies in length among cultivars, it is conceivable that, in the absence of pollinating insects, the long style would prevent pollen from the anther reaching the stigma, and fruit setting would be prevented or reduced.

POLLEN

The pollen grains of pepper are elliptic, three-segmented and light yellow in colour. There are somewhere between 11,000 to 18,000 pollen grains in a single anther (Hirose, 1957). A normal fertile flower may contain 1–1.5 mg of pollen (Quagliotti, 1979). The average length of dry pollen grains is 20.4–40.3 nm, with a dehisced pollen grain having a diameter of 17–38 μm, again depending on the cultivar. When the pollen grains are placed on the stigma and they become imbibed, their size increases to nearly twice the original.

Air temperature can have a large effect on pollen formation and viability. The optimal temperature for pollen germination is 20–25°C. Pollen formation is harmed when the temperature is above 30°C. Cochran (1938) reported that temperatures above 30°C 15 days prior to anthesis would cause sterile pollen. Hirose (1957) observed that in tabasco (*C. frutescens*), the dehiscence of the anthers occurred relatively late in the morning between 10:00 a.m. and noon. Pollen can be stored for later use in breeding programmes. At 0°C pollen can be stored for 5–6 days; if the humidity of the storage area is lowered to 56%, the pollen storage time can be extended to 180 days (Ghatnekar and Kulkarni,

1978). Pollen tube growth from the stigma to the egg has been reported to take anywhere from 6 to 42 h.

Cytoplasmic male-sterility (CMS) is found in *Capsicum* (Peterson, 1958). When Horner and Rogers (1974) examined the morphology of CMS, they found that the tapetal cells of the anthers become highly vacuolate during meiosis, and remain appressed to the microsporocytes. The locular cavity, which is a general feature of fertile anthers, is not formed. Meiosis in the microspore mother cells proceeds normally, and the primexine develops in tetrads. Further development of the microspores is arrested, and the tetrads collapse even while they are enclosed in the callose wall. The use of male sterility in pepper breeding is discussed in Chapter 5.

CALYX/PEDICEL

With *C. annuum*, the calyx is campanulate and has 5–7 dentate 'teeth', ribbed, and about 2 mm long. This is not true for other *Capsicum* species. A useful taxonomic characteristic of the calyx is the annular constriction found only in the *C. chinense* species. The calyx usually encloses the base of the flower. The calyx is persistent in most domesticated species; however, the wild species all have a non-persistent calyx. With wild species the calyx separates readily from the fruit. This trait is advantageous for seed dispersal. Frugivorous birds, which are seed vectors in nature, are able to remove the fruit easily. The gene for the non-persistent trait is the soft pedicel (*Sp*) and is dominant. Through domestication, this trait has been selected against, thus giving fruit that tends to stay attached to the plant.

The fruit pedicel length varies in the different pod types and ranges from 10 to 20 mm long. The small-sized fruits generally have longer pedicels. *C. baccatum* var. *pendulum* has a characteristically long pedicel. In dried fruits of the New Mexican pod types, the calyx and pedicel account for 7% of the total weight.

FRUIT DEVELOPMENT

As described in Chapter 2, there is extensive diversity in fruit shapes, sizes and colour. Among the different pod types fruit length can vary from less than 1 cm to, as stated in the Guinness Book of Records, 32.5 cm. Fruit growth is dependent on ovule growth, whether it is fertilized or not. The fruit is usually seeded, but seedless, parthenocarpic forms do exist. The seeds affect the development and growth of the fruit (Marcelis and Hofman-Eijer, 1997). Seed number affects the fruit's growth rate rather than its growing period. When seed number increases in a fruit there is an inhibitory effect on fruit set and growth of later-developing fruits.

The pod may have two or more locules each divided by a central placenta. The placenta has the vesicles for oleoresin and capsaicinoid production (see Figs 3.10 and 3.11). However, its most important role is to provide nourishment for seed development. The 'walls' or pericarp of the pod consists of epidermal cells in regular order with a thick-grooved cuticle. The cuticle thickness differs among the pod types. Several rows of collenchymatously thickened, beaded cells constitute the hypodermis. The mesocarp is formed by thick-walled beaded cells, while the inner mesophyll cells are thin-walled ground parenchyma and fibrovascular bundles. The vascular bundles of the pericarp consist of xylem tissue with spiral vessels, and phloem tissue. 'Giant cells' are situated between the mesocarp and the endocarp tissues. These giant cells cause the numerous 'blisters' seen on the inner surface of the pericarp.

The pod shape is based on cell division, which takes place at the preanthesis stage. Fruit size is determined by elongation during anthesis and postanthesis. The fruit growth zone, especially in the New Mexican type, is situated mainly at the tip of the fruit. This is why blossom end rot is found at the tip. For the bell pod types, the growth is more evenly distributed over the whole length of the pod. As with many fruits, the pepper fruit growth curve is a simple sigmoid type.

The time from anthesis to fully grown fruit varies considerably among the different pod types. Maturity depends on cultivar and the environmental conditions during maturation. Temperature can affect fruit development indirectly because temperature does affect vegetative growth. Thus if assimilate from the leaves is limited there can be an effect on the pods. Pods are harvested at the immature and mature stages. The green stage is horticulturally ripe, but physiologically immature, meaning that if the fruit are picked, they are incapable of ripening normally. The fruits are characterized as non-climacteric in ripening (Lownds *et al.*, 1993).

Parthenocarpy can be induced in peppers by genetics (Curtis and Scarchuk, 1948), non-optimum air temperatures and plant growth regulators. Parthenocarpic fruit is abundant when peppers are subjected to cool day temperatures or high night temperatures. Notwithstanding, all parthenocarpic fruit have greater deformation in shape and size than normally formed fruits. Sometimes the development of extra fruit-like bodies inside the fruit is seen.

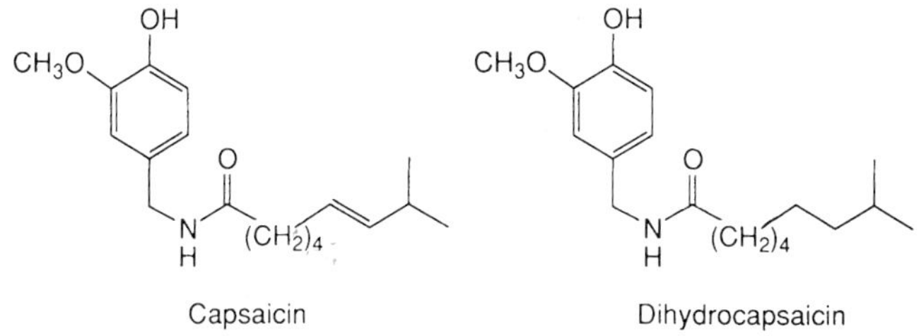

Fig. 3.10. The structures of capsaicin and dihydrocapsaicin.

Histological studies show that the style of the normal flower and that of the internal fruit-like body are similar in some tissues, but the fruit-like bodies fail to develop a stylar canal. Both the main fruit and the internal bodies have an epidermis, large parenchyma cells, chloroplasts and vascular bundles. The main fruit wall contains several layers of cells that constitute the epicarp. The internal fruit-like bodies lack these layers of cells.

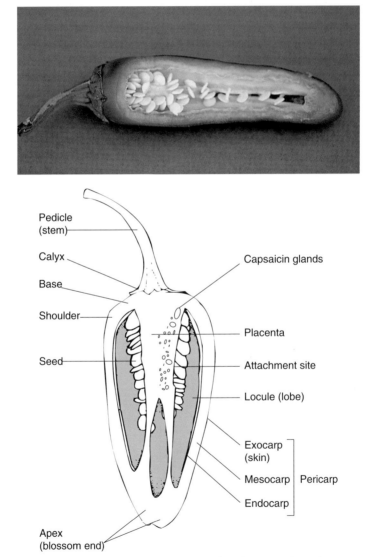

Fig. 3.11.   Pepper fruit illustrating the vein on the placenta where the capsaicinoids are concentrated (pungency).

During ripening of the pod, chlorophyll disappears and the carotenoid content increases. The biosynthetic pattern of the carotenoids in pepper is very different from that of tomatoes. The type and amounts of carotenoids differ significantly among the different pod types. The carotenoid content is controlled by the genotype of the plant and the environment where the plant is grown. Chromoplasts synthesize and accumulate large amounts of carotenoid pigments. Carotenoids are responsible for the yellow, orange and red colour of the fruits. The main difference between yellow and red fruit colour of physiologically ripe fruits is that yellow fruits contain lutein and violaxanthin as the major carotenoids, together with other similar xanthophylls, while in the red fruits lutein is completely absent. The major carotenoids are capsanthin and capsorubin.

Production practices have a large effect on the fruit produced. Plant density, environmental conditions and the cultivar chosen will all contribute to the final qualities of the fruit. Depending on the production method (greenhouse or field) and the pod type grown, artificial inputs can aid in fruit development. Supplemental lighting, plant protection, e.g. plastic tunnels, and increased nutrients, can aid in producing a high quality fruit.

SEEDS

INTRODUCTION

Seeds can be viewed as the most important part of the production of peppers. Healthy, normal seeds (see Fig. 4.1) will produce a high yielding, excellent quality crop. Because of the importance of seeds, much research has been undertaken in this area. This necessitated a separate chapter covering the extensive information published about pepper seeds.

POLLINATION

Pepper plants are considered to be a self-pollinating crop (Allard, 1960). However, the rates of out-crossing (2–90%) recorded by several investigators argue that *Capsicum* should be considered a facultative cross-pollinating species in field research (Odland and Porter, 1941; Franceschetti, 1971; Tanksley, 1984a). Even though the amount of out-crossing varied among the

Fig. 4.1. Seeds of *Capsicum annuum* (left) and *Capsicum pubescens* (right).

investigations, it was nevertheless sufficient to impede progress in breeding programmes (Odland and Porter, 1941; Tanksley, 1984a). The out-crossing is associated with natural insect pollinators, not rain or wind (Odland and Porter, 1941; Tanksley, 1984a). The amount of cross-pollination has an effect not only on the precautions needed for seed production, but also on the breeding methodologies used by the plant breeder. Natural pollinators such as insects must be excluded to ensure self-pollination. Peppers are cross-pollinated to such a considerable extent that those undertaking seed production activities must be cognizant of this. Ants have frequently been mentioned as pollinators of pepper. Their type of activity, the lack of a dense coat of hairs on their body and their limited number in relation to the blossoms present in a commercial planting, cast doubt on their ability to cross-pollinate peppers. Honey bees and solitary bees are much more likely to cross-pollinate peppers. To produce large amounts of genetically pure seed, seed certification programmes employ isolation as the control mechanism (NMCIA, 1992). These isolation requirements range from 1.6 km for the Foundation class of seed to 0.4 km for the Certified class of seed.

In breeding programmes, numerous breeding lines and plants must be isolated during seed production. Thus, space for isolation becomes limiting. To ensure self-pollination, a simple and effective plant isolation cage was developed (Bosland, 1993). The cage consists of nylon fabric draped on a frame constructed of conduit piping (see Fig. 4.2). Green and white coloured fabric have been successfully used. With either colour no effects on plant growth, fruit set or seed production have been observed under southern New Mexico climate conditions. The fabric mesh size has a count of 20 holes by 16 holes/2.5 cm^{-2}. This mesh effectively prevents the entry of pollinating insects into the cage. It is not necessary to anchor the cages with soil. Even though New Mexico has strong winds the fabric is heavy enough to lie flat on the ground without soil being mounded on the edge. At the end of the season, the cage is washed and stored for the next season.

While the cage excludes cross-pollinating insects, it simultaneously prohibits the natural beneficial insects that control aphids (*Myzus* spp.), the major pest found inside the cage. Chemical insecticides sprayed through the fabric will control the aphids. However, a biocontrol measure, the use of ladybirds (*Hippodamia convergens* Guérin) has been effective in controlling aphids (Votava and Bosland, 1996). For each isolation cage, 75–100 ladybirds are placed inside at the time the cage is placed over the pepper plants. If caterpillars are encountered under the cages during the growing season a treatment of *Bacillus thuringiensis* is used with no harm to the ladybirds.

Fig. 4.2.   Cages used to prevent out-crossing among pepper accessions.

HYBRIDIZATION

A plant breeder transfers the pollen from the anther of one flower to the stigma of another to make controlled hybridization. It is also important to control pollination when producing F_1 hybrid seed. Hybrid seed is the seed of an F_1 generation sold for commercial production of the crop. Peppers grown from hybrid seed are highly uniform and usually higher yielding. The production of hybrid seed is discussed in detail in Chapter 5.

DESCRIPTION

The seed develops from a campylotropous ovule, meaning that the ovule is curved so that the micropyle is located near the base. Within the pod the seeds are attached to the placenta in close rows, principally near the calyx end of the pod. Pepper seeds are described as flat and disc-like in shape, with a deep chalazal depression.

Pepper embryos are surrounded by a well-defined endosperm in the seed which makes up the bulk of the food reserves for the embryo and the young seedling (see Fig. 4.3). The endosperm lies directly in front of the radicle and is

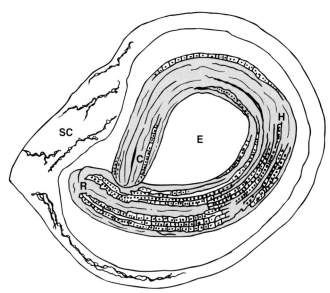

Fig. 4.3. Diagram of a longitudinal cross-section of a pepper seed showing: C, cotyledons; E, endosperm; H, hypocotyl; R, radicle; and SC, seed coat.

seven to nine cells thick (Watkins *et al.*, 1985). *C. annuum* seeds have mainly protein and lipid as storage reserves (Chen and Lott, 1992). Endosperm cells bordered by the internal epidermis are angular in shape, have slightly thickened walls and include oil and aleurone granules of crystalloid content.

The endosperm provides a supplemental nutrient source for early seedling growth. Watkins *et al.* (1985) have shown that the thickened cell walls of *C. annuum* endosperm probably act as a store of mannan-containing polysaccharides. With pepper, the external appearance of the endosperm changes one day before radicle emergence, when the endosperm in front of the radicle enlarges and protrudes outward (Watkins *et al.*, 1985). This change is accompanied by a loss of integrity in the endosperm and a reduction in thickness directly in front of the radicle, but not in other regions of the endosperm. These changes confirm that the endosperm serves as an additional food source for the developing pepper seedling.

A disorder of pepper seeds called 'fish-mouth' occurs when the seeds are harvested immaturely and the endosperm has not fully developed. The seed has a characteristic fish-mouth appearance, hence the name.

The cells of the outermost layer of the endosperm next to the seed coat and the embryo protoderm cells are covered by a cuticle. The growing point or plumule is between the cotyledons. As with all dicotyledonous plants, there are two cotyledons that are tightly appressed. Pepper cultivars differ in their shape and size of cotyledons. After germination, the typical cotyledon is green in colour and is wider in the middle, then narrows towards the apex and the base part. There are some cultivars that have purple or variegated cotyledons.

The seed is covered by a parchment-like seed coat. As with other solanaceous crops, the seed coat is derived from a single integument. The seed coat is usually smooth, but some can be slightly rough/subscabrous. *C. pubescens* and *C. lanceolatum* have a detailed reticulated pattern on the seed surface. The seed coat does not cause any mechanical restriction to germination (Watkins and Cantliffe, 1983b).

Seed colour is straw yellow, tan or black. As seeds age and lose viability they can become brown. Seed size is dependent on the variety and the growing conditions, usually larger fruits will have larger seeds. Most seed falls in the range of 2.5–6.5 mm in length and 0.5–5 mm wide. An average *C. annuum* seed is about 1 mm thick, 5.3 mm long and 4.3 mm wide with a surface area of 33 mm^2 (Chen and Lott, 1992). A thousand seeds weigh 5–7 g. Seeds account for approximately 20% of the dry weight of peppers.

Seed size affects the uniformity of pepper plants. Cochran (1974) found that seeds having a diameter of 3.5–4.2 mm and having a 20 seed weight of 7.2–8.1 mg emerged 2 days earlier, had a significantly better stand and produced overall better plants than small seeds measuring less than 3.0 mm and weighting less than 5.9 mg. The small seeds failed to produce transplants meeting the minimum size requirements for transplants.

DORMANCY AND GERMINATION

Freshly harvested seeds of *Capsicum* can exhibit dormancy. It is recommended that an after-ripening period of about 6 weeks is required at room temperature to remove dormancy (Randle and Homna, 1981). However, the authors have taken seed directly from red ripe New Mexican and jalapeño fruits from the field and planted them directly in the greenhouse with excellent success in germination.

When temperature effects were examined with non-dormant seeds, all *Capsicum* species germinated well when tested over a constant temperature range of 15°C and 30°C, while *C. baccatum* non-dormant seeds germinated fully at 10°C and 13°C (Randle and Homna, 1980). Alternating temperature regimes 15/27°C or 15/30°C promoted the germination of dormant seeds of *C. annuum*, *C. baccatum*, *C. chinense*, *C. frutescens* and *C. pubescens* quite substantially (Gerson and Homna, 1978). The best temperature regime for dormant seeds is 30/15°C (16 h/8 h) for 14 days.

It appears that no special requirements for light are necessary for pepper seed germination. Fluorescent light sources used in germination cabinets do not inhibit *Capsicum* seed germination. However, they do not promote germination either. Therefore, the presence or absence of light is not a factor in *Capsicum* seed germination.

Peppers have a prolonged germination period and an optimum germination temperature of about 30°C. The rate of germination and emergence is markedly reduced at temperatures in the range of 15–20°C. Hastening the germination and emergence of pepper seed, especially at suboptimal temperatures, would be of significant value in the production of greenhouse-raised plants. Various seed treatments have been suggested to improve seed germination and seedling emergence in pepper.

SEED TREATMENTS

Successful dormancy-breaking treatments have included potassium nitrate and gibberellic acid (GA$_3$). A solution of 2 g l^{-1} of potassium nitrate was successful in eliminating dormancy of *C. annuum* seeds. Seeds are soaked for 4 h before planting. GA$_3$ solutions of 100 ppm and 1000 ppm were also successful in overcoming dormancy (Watkins *et al.*, 1985). The solutions can be used daily with watering until the seeds germinate.

Treatment of seeds with sodium hypochlorite is a well established procedure to disinfect seeds of pathogen contamination (Goldberg, 1995). Sodium hypochlorite has been reported to promote germination (Fieldhouse and Sasser, 1975). The effect of sodium hypochlorite on seed germination is dependent on seed age, treatment time and temperature (Khah and Passam, 1992). Fresh seeds are more sensitive to the treatment than seed stored for 10 months at ambient temperature. Khah and Passam (1992) recommended that pepper seeds be treated with a 3% solution at 10–25°C for up to 20 min. This treatment should surface sterilize the seeds without inhibiting germination.

SEED PRIMING

Investigations of techniques to improve pepper germination and emergence are numerous and somewhat conflicting. One seed treatment method that has proven successful in increasing pepper seed vigour in most cases is osmoconditioning, or priming. Seed priming is the process of soaking seed in an osmotic solution. Priming enables the seed to germinate and emerge faster at suboptimal temperatures.

In general, seed priming improves germination percentage and rate, emergence, seedling growth and uniformity, and yields of peppers. Studies involving seed priming have primarily focused on sweet pepper types and have produced variable results. Treating pepper seed by soaking in potassium salt solutions has been shown to improve germination and emergence in soil temperatures of 10–14°C (Gerson and Homna, 1978). Priming seeds in solutions of potassium nitrate (KNO$_3$) (Bradford *et al.*, 1990) or in polyethylene glycol (PEG) (Yaklich and Orzolek, 1977) also enhanced the rate of germination. Sundstrom and

Edwards (1989) reported an increased rate of germination when jalapeño and tabasco peppers were primed in 3.0% or 2.75% KNO_3 solutions, respectively. Priming seed of jalapeño and tabasco in 3.0% KNO_3 for 144 h and -4 bar polyethylene glycol-6000(PEG) for 120 h enhanced germination rates when tested in the temperature range of 5–35°C (Rivas *et al.*, 1984). They also found that surface drying seed following priming retarded germination rates of both jalapeño and tabasco over all temperatures tested as compared with primed seed not surface dried. In contrast, Ghate and Phatak (1982) reported a significant decrease in germination rate when pepper seeds were primed with K_2HPO_4 plus $(NH_4)_2HPO_4$ solution.

Low temperature germination and emergence of pepper seed can be accelerated by priming at 20°C. The stimulating effect of salt solutions on germination was most marked at 12.5°C. At this temperature, time to 50% germination was up to 14 days less for treated seed than for untreated seed. Imbibed seed emerged up to 6 days earlier than untreated seed (O'Sullivan and Bouw, 1984). The results of O'Sullivan and Bouw (1984) showed that imbibing pepper seed in salt solutions can accelerate germination and emergence at suboptimal temperatures. For germination at moderately low temperatures, the optimum conditions were treatment in a 0.75% salt solution for 10 days at 20°C followed by surface drying, sufficient only to allow proper handling of the seed at planting.

The main function of the salt is to maintain an osmotic potential sufficient to prevent the seed from germinating while permitting enough moisture to enter the seed to allow completion of the early metabolic steps in germination. Smith and Cobb (1991) found that a specific ion or salt was not essential in priming pepper seeds, and that effective priming is strongly dependent on both the osmotic potential of the priming solution and the duration of the treatment.

Seed priming can markedly increase the germination rate of pepper seed, but has not always been accompanied by improvements in field emergence rates or percentages (Yaklich and Orzolek, 1977; Ghate and Phatak, 1982). Rivas *et al.* (1984) found that despite positive responses to seed priming in laboratory tests, emergence of tabasco pepper in the field was not improved by priming. However, some field emergence trials did find a more rapid and a higher percentage of emergence from primed seed (Martinez and Aljaro, 1987). Furthermore, optimal priming conditions may vary among cultivars and even seed lots of a given species (Bradford *et al.*, 1990). Germination and emergence of some types of hot pepper seedlings, even under favourable conditions, often requires up to 2 weeks.

The control of germination at suboptimal temperature may be governed by inhibitors or promoters. Watkins and Cantliffe (1983a) did not find any evidence of a leachable or extractable germination inhibitor being activated or formed during exposure to low temperature. Auxin and kinetin applications did not alter germination rates, but gibberellins (GA_3 and GA_{4+7}) increased germination rates. GA_{4+7} was slightly more effective than GA_3 in stimulating a germination rate increase. AMO-1618, an inhibitor of gibberellin synthesis, delayed germination of pepper seeds. It appeared from their study (Watkins and Cantliffe, 1983a) that low levels of gibberellin may be synthesized prior to radicle protrusion in pepper seeds.

Kanchan (1973) reported that soaking pepper seed for 5–10 h in GA$_3$ solutions stimulated germination. Seed germination of pepper in aerated water columns was accelerated and germination uniformity improved by using GA$_3$ at 6 μg mg^{-1} seed with 50–75 mg seed ml^{-1} of solution (Sosa-Coronel and Motes, 1982). Higher GA$_3$ rates in the aerated columns reduced germination percentage in some cultivars tested.

Stoffella *et al.* (1988) have described the sequence of early root development in bell peppers as a rapid taproot elongation from radicle protrusion until cotyledons are fully developed, emergence of basal and lateral roots when cotyledons are fully developed and an increase in lateral and basal root numbers, with a simultaneous reduction in taproot growth rate. Priming seeds caused no beneficial or deleterious effect in seedling root morphology.

Plug-mix, fluid drilling and gel-mix delivery systems have been developed as carriers of non-treated, primed or pregerminated seeds for commercial field conditions. Because pregerminated seeds emerge more rapidly than dry seed, the difference in speed of emergence is accentuated at lower temperatures, which makes the practice of using primed seed particularly attractive for direct-seeding in the field. The optimum radicle length is 2–4 mm because a longer radicle would be damaged during planting.

It is evident, however, that if the pregerminated seeds cannot be sown immediately they must be carefully stored to prevent debilitation. Successful storage of pregerminated seed is dependent upon the temperature being low enough to temporarily limit radicle growth and respiration, but not so low as to cause chilling injury. When pregerminated pepper seeds were tested for sensitivity to low temperatures, it was found that at 5°C seed could be stored for 21 days without a reduction in percentage emergence when compared with fresh pregerminated seed (Irwin and Price, 1981). Storage of pregerminated pepper seed at 5°C has been shown to be an effective means of temporarily halting radicle growth in the event of delayed planting.

COATED SEED

Coated seed is used to facilitate precision planting. Coated seeds are usually slower to germinate, but have no effect on overall growth or yield of the crop. An additional advantage to seed coating is that it enables the placement of beneficial organisms or chemicals in close proximity to the seed. Clays have been used as the primary coating material, but bell pepper seed lose their ability to geminate properly when coated with clay (Sachs *et al.*, 1981). The germination rate of sand-coated pepper seed was faster than clay-coated seed, but was slower than untreated seed (Sachs *et al.*, 1982). It was suggested that oxygen may be limiting for the germinating seed when coated.

CHEMICAL COMPOSITION

Bush (1936) has reported on the chemical composition of pepper seeds. He sampled from a 1 year accumulation of 16 tons of seeds in California. He found that the seeds contained 26.10% oil, 6.25% moisture and 67.65% dried extracted meal. When the meal was analysed, it was found to contain 28.92% protein, 29.10% fibre, 5.61% ash and 36.37% N-free extract (carbohydrates). The properties of the oil were a specific gravity of 0.918, a refractive index at 25°C of 1.4738, colour (2.54 cm column, Lovibond) scored 100 yellow -46 red, an acid number of 2.18, a Hanus iodine number of 133.5 and an acetyl number of 7.0. The saponification number was 192.0 with an unsaponifiable matter of 1.7%. The melting point of separated fatty acids was 21.1°C. These values are similar to those found for most edible oils.

A study by Reddy and Sarojini (1987) using Indian-type chile peppers reported that 60% of the dry weight of the chile peppers was seeds. Other pepper pod types, for example bell pepper or New Mexican, will have less seed as a percentage of the dry weight. The oil content of Indian chile pepper seeds was 12–26%, depending on the cultivar tested. The oil was rich in unsaturated fatty acids: 70.6% linoleic and 10.9% oleic with small amounts of myristic (0.2%), palmitic (16.3%) and stearic (2.2%). Reddy and Sarojini (1987) recommended that chile oil be used as a cooking oil because it would supply essential fatty acids and contribute to the flavour of the dish being prepared.

Itoh and his research group (Itoh *et al.*, 1977, 1978, 1979; Matsumoto *et al.*, 1983) have identified 17 4α-methylsterols of which four were novel sterols found only in *Capsicum*. They reported the occurrence of two 24(*E*)-ethylidene sterols, fucosterol and 28-isocitrostadienol in the unsaponifiable materials of *Capsicum annuum* seed. The research group has also shown that Z-isomers of the above two 24-ethylidene sterols, 28-isofucosterol and citrostadienol are also present in *Capsicum annuum* oil.

SEED YIELD

Harrington (1960) found that nutrient-deficient pepper plants produced lower seed yields than did control plants. He also demonstrated that phosphorus nutrition of parent plants failed to influence seed performance of the progeny. When he studied potassium-deficient plants, they gave a higher proportion of abnormal seeds with dark-coloured embryos and seed coats. Both normal and abnormal seeds from such plants had a lower percentage of germination than control seeds, and their viability declined more rapidly in storage. Gill *et al.*

(1974) reported that an increase in applied nitrogen from 0 to 370 kg ha^{-1} did not produce any proportional increase in seed yield. Payero *et al.* (1990) applied solubilized ammonium nitrate (NH$_4$NO$_3$) through a trickle-irrigation system to ensure uniform and timely applications of nitrogen. They found the maximum red fruit production resulted from the highest nitrogen application level (240 kg ha^{-1}). When seed parameters were measured as an effect of the nitrogen fertilization, final germination percentage, seedling root length and weight, and field emergence were unaffected by any of the nitrogen treatments. Payero *et al.* suggested different nitrogen management strategies for seed production and for fruit production.

5

GENETICS, PLANT BREEDING AND BIOTECHNOLOGY

INTRODUCTION

The earliest pepper breeders were the indigenous peoples of the Americas. The native Americans selected and developed many of the pod types of peppers we know today. These include jalapeño, serrano, pasilla and ancho, to name a few. Today's pepper breeders are faced with the task of assembling into a cultivar the superior genetic elements necessary for increased yield, protection against production hazards and improved quality. The specific goals of these breeders are as numerous as the many types of peppers and their subsequent uses.

Each type of pepper must conform to its own unique set of characteristics in order to be commercially acceptable. A plant breeder working with bell peppers, for example, will have different objectives from a breeder concerned with paprika cultivar development. In addition, the end use of each type of pepper must be considered. Green, ripe New Mexican-type pepper destined for fresh market sale may have very different horticultural requirements from New Mexican-type pepper grown for red powder.

Some characteristics are difficult to manipulate. Pungency is one such characteristic which is hard to stabilize. Growers, processors and consumers have very specific demands in terms of expected pungency for different types of pepper or pepper products. Unfortunately, pungency is a character that is severely altered by environmental conditions during production (Harvell and Bosland, 1997), such that stable genotypes must be found first in order to attempt to breed for a cultivar that has a stable pungency.

Breeders, therefore, are faced with a plethora of unique objectives, and in order to achieve those objectives, they may use quite a few different breeding methods. First, however, an understanding of pepper genetics and reproductive behaviour is required.

GENETICS

The nuclear DNA content of various *Capsicum* species, as determined by flow cytometry, ranges from 7.65 to 9.72 pg per nucleus for *C. annuum* and *C. pubescens*, respectively (Belletti *et al.*, 1995). Most species of pepper are diploid, with 24 chromosomes ($2n = 2x = 24$), and have one or two pairs of acrocentric chromosomes with ten or eleven pairs of metacentric or submetacentric chromosomes (Lanteri and Pickersgill, 1993). Interestingly, *C. annuum* and *C. chinense* differ from one another by means of two chromosomal interchanges (Tanksley and Iglesias-Olivas, 1984; Lanteri and Pickersgill, 1993).

A list of known genes can be very useful to pepper breeders, especially if a collection of germplasm that contains representative specimens is available. In 1965, Lippert *et al.* produced a gene list for pepper. The list included 50 genes and a standardization of rules for naming and symbolizing. In 1994, Daskalov and Poulos produced an updated gene list. Likewise, the authors described a protocol for names and symbols, as follows: The *C. annuum* cultivars 'California Wonder' and 'Doux Lond des Landes' are the normal or 'wild' type standard genotypes. Normal or wild type alleles are shown with the superscript plus sign (+), e.g. A^+. A gene symbol is assigned by a maximum of three letters that best abbreviate the gene name. The gene symbol and name should begin with the same letter. If the gene is dominant the symbol will begin with a capital letter, e.g. *Dms*. If the gene is recessive, then all lower case letters will be used. A gene best describes the phenotype. Additional loci are given a number in addition to the gene symbol when more than one locus affects or shows the same trait, e.g. *c1, c2*. Allelic and complementation tests should be used to validate assignment of additional loci, unless the trait is found in more than one species where barriers to interspecific hybridization make genetic studies difficult. Genes from other species should be clearly indicated in the description of the trait. Numbers are written on the same line as the gene symbol. Multiple alleles at a locus are noted by superscripts. Capital letter or Arabic number superscripts are used for dominant alleles; lower case letters for recessive alleles. Recessive mutants for which dominant alleles are later discovered are noted by the recessive gene symbol (lower case) with a capital letter superscript, e.g. vg^M, vg^V, vg^H. Order of dominance of alleles should be put in the description using the '>' symbol, e.g. $L4 > L3 > L2 > L1 > L^+$. Assignment of multiple alleles should be validated by allelic tests. The entire gene symbol should be italicized or underlined. New symbols should never be assigned in opposite case to an already assigned symbol nor should the same trait be described by different symbols.

At the same time the Capsicum and Eggplant Newsletter Committee for Capsicum Gene Nomenclature published additional rules for gene nomenclature. An important aspect of this document was the call for a seed sample of the named and accepted gene stock. All genetic samples are to be deposited in the Capsicum Genetics Cooperative at New Mexico State University, USA. Duplicate samples must be maintained by the originator or at a separate location (CENL, 1994).

Cytogenetic studies examining chromosomal structure have been carried out by several researchers. Pickersgill (1971) examined karyotypic differences as well as geographical distribution, crossability and archaeological data in order to clarify the relationships between cultivated and wild species of *Capsicum*. Cultivated *C. annuum* contained two pairs of acrocentric chromosomes with satellites on one or both pairs, whereas weedy forms contained only one or rarely two acrocentric pairs. The weedy forms had much more variable karyotypes than the cultivated forms, leading Pickersgill to suggest that cultivated forms were domesticated relatively recently from a small population of wild *C. annuum* and have therefore gone through a genetic bottleneck. Variability in karyotype was also greater in wild forms of *C. baccatum*, *C. chinense* and *C. frutescens*. Moscone (1990) performed a karyotypic analysis of two populations of *C. chacoense*. In 1993, Moscone *et al.* utilized Giemsa C-banding of six *Capsicum* species in order to clarify taxonomic grouping. Lanteri and Pickersgill (1993) examined the meiotic chromosomes of an F_1 hybrid between *C. annuum* and *C. chinense* and confirmed that the two accessions differed by two chromosomal interchanges involving three pairs of chromosomes. In 1995, Moscone *et al.* used silver staining of active nucleolar organizing regions of nine *Capsicum* species in order to provide additional markers for chromosome identification.

FLOWERS AND MODE OF REPRODUCTION

Understanding flower morphology is a necessity for the pepper breeder. Flower morphology is discussed in greater detail in Chapter 3. The salient facts of flower morphology important to pepper breeders are addressed here. Pepper flowers are complete, that is they have a calyx, corolla, and male and female sex organs. Most species of pepper are self-compatible. Self-incompatibility has been reported in *C. cardenasii* and in some accessions of *C. pubescens* (Yaqub and Smith, 1971). Mating among siblings is required to produce viable seed within these accessions. Peppers exhibit no inbreeding depression. All species are protogynous and can cross-pollinate. The stigma is positioned slightly below, level with the anthers or exerted beyond, in which case the chances for cross-pollination are greater. Studies have shown that cross-pollination can range from 2 to 90% (Pickersgill, 1997). Therefore, pepper breeders and seed producers must use caution to prevent uncontrolled cross-pollination (Bosland, 1993).

MAKING CONTROLLED HYBRIDIZATION

In order to make controlled hybridizations, a plant breeder must transfer the pollen from the anther of one plant to the stigma of another. To prevent self-pollination, unopened flower buds are chosen. Using alcohol-sterilized forceps, the petals are carefully removed to expose the reproductive organs. The flower is then emasculated by removing all anthers. The stigma is then examined for any pollen 'contamination' before making the controlled cross. Pollen is transferred from the open flower of the male, or pollen donor, using a small paint brush, bee stick or by direct contact with the anthers. The pollen is placed on to the stigma of the emasculated female plant. The cross is labelled and, after the fruit ripens, the seed is collected.

BREEDING METHODS

A variety of breeding methods can be used to produce new pepper cultivars. The methods used are determined by the breeder to best fit the goals of the breeding programme. Modern breeding efforts have focused predominantly on sweet pepper cultivar development. Development of pungent cultivars has occurred to a relatively lesser extent (IBPGR, 1983).

Mass Selection

The first pepper breeders, the indigenous people of tropical America, domesticated five different species of *Capsicum* in five different domestication events and locations. These early pepper breeders used the technique of mass selection whereby seeds of the best plants were saved for the next growing season. Plants best adapted to a specific geographical area were selected and thus thousands of landraces were developed. After Columbus's voyage, peppers quickly spread around the world and, again, landraces were formed that filled specific cultural and environmental niches. Today, these landraces are sources of genetically diverse germplasm.

Pedigree

The pedigree method involves keeping records of the matings and their progeny. This includes making single plant selections and self-pollination. The pedigree of subsequent selfed generations is recorded in combination with selection for desired traits. The system produces homogenous lines (Fehr, 1993). 'NuMex Joe E. Parker' is an example of a cultivar developed by this method (Bosland *et al.*, 1993). It originated as a single plant selection from a field planted to an open-pollinated population of 'New Mexico 6-4'. A pedigree selection protocol was followed for three generations under greenhouse conditions. Each line was then evaluated for more than 25 horticultural traits.

Backcross

The backcross method uses a successful cultivar as a recurrent parent following an initial cross between that successful cultivar and a separate individual that serves as a donor parent for a given desired character. Following successive backcrosses to the recurrent parent, the breeder arrives at a cultivar identical to the recurrent parent but containing the additional desired trait from the donor parent. The backcross method is successful only when attempting to introgress a trait controlled by one or at most a few genes. An example of a successful pepper cultivar developed via the backcross method is the 'Greenleaf Tabasco'. In 1950, Walter Greenleaf crossed tabasco pepper with an accession of *C. chinense* resistant to tobacco etch virus (TEV). At the third backcross, Greenleaf made a cross to a second TEV resistant *C. chinense* donor parent. The recessive mode of inheritance for TEV resistance required that alternate generations be screened. In 1970, after several backcrosses, 'Greenleaf Tabasco' was introduced (Greenleaf, 1986).

Recurrent Selection

Recurrent selection is a breeding method that involves selecting individual plants from a population followed by intercrossing to form a new population. The intention of this system is to produce transgressive segregants, i.e. individuals that contain greater resistance as a result of genetic segregation resulting from crosses between parents that already have some degree of resistance (Palloix *et al.*, 1990a). Current efforts that utilize this method for pepper breeding have focused on developing pepper cultivars resistant to *Verticillium dahliae* and *Phytophthora capsici* (Palloix *et al.*, 1990a,b).

Mutation Breeding

Mutation breeding is a means by which mutations are generated in peppers to improve economically important traits or to eliminate deleterious traits. It is not a major breeding method but may be a means of producing novel mutants of interest. Mutations can be induced chemically or by ionizing radiation. Mutations of interest may then be incorporated into commercially acceptable lines by traditional breeding methods such as the ones described above. Bhargava and Umalkar (1989) used both gamma radiation and chemical mutagens to produce an array of pericarp mutations. Alcantara *et al.* (1996) described optimal conditions necessary for seed mutagenesis in *C. annuum* using the mutagen ethyl methanesulphonate (EMS).

HYBRID SEED PRODUCTION

Hybrid seed is the seed of an F_1 generation sold for commercial production of the crop. Peppers grown from hybrid seed are highly uniform and usually higher yielding. Several systems to produce hybrid seed are possible, including the use of genetic male sterile plants and cytoplasmic male sterile plants. Unfortunately, the production of today's pepper hybrids commonly relies on making crosses between the two parents by hand; a very labour intensive and expensive process.

Genetic Male Sterility

Genetic male sterility is one means by which hybrid seed may be produced. The sterile plants are used as the female parent of a hybrid cross. The male sterile characteristic is often inherited as a single recessive gene, *ms*. The use of genetic male sterility is limited in hybrid seed production due to the inefficiency of producing and maintaining a population of male sterile plants. In order to produce more male sterile plants one must cross a fertile plant heterozygous for the male sterile trait to the male sterile plant, and then only half the progeny from this cross will be male sterile.

Male sterility in peppers has been extensively studied by Shifriss and others. In 1969, Shifriss and Frankel found a genetic male sterile plant in a population of the cultivar, 'All Big', bell pepper. They discovered that the character was determined by a single recessive gene. The authors also noted that male sterility was accompanied by the development of parthenocarpic fruit throughout the growing season. In 1972, Shifriss and Rylsky described the discovery of a second gene encoding genetic male sterility. The second source came from a population of 'California Wonder' bell peppers. Again, the character was inherited as a single recessive gene. The authors suggested that the genetic male sterile plant described by Shifriss and Frankel in 1969 should be called *ms1*, and the second non-allelic gene *ms2*.

In an attempt to increase the ratio of male sterile plants in a population, Shifriss and Pilovsky (1993) crossed two isogenic lines that differed for male sterility genes. The intention of this digenic cross was to a produce a system in which a male sterile plant contained both previously described male sterile genes, i.e. *ms1* and *ms2*. This plant was then crossed to a fertile plant that was heterozygous for both genes, i.e. *ms1ms1ms2ms2*×*Ms1ms1Ms2ms2*. The resultant progeny from such a cross was segregated in a ratio of three male sterile plants to one fertile plant. The implication for such a cross was that only a quarter of plants would have to be removed from a seed production field. Unfortunately, the procedure required both parents to be maintained asexually and protected from viral contamination (Shifriss, 1997). In general, genetic male sterility systems for hybrid seed production in pepper have not been used to any significant level owing to the production of a high percentage of non-hybrid seed and because of the labour-intensive nature of the system (Daskalov and Mihailov, 1988).

Cytoplasmic Male Sterility

Cytoplasmic male sterility (CMS) is another means by which hybrids may be produced. The advantage of a CMS system is that a population of sterile plants can be generated in which all the offspring are sterile. Sterility results from an interaction of nuclear and cytoplasmic factors. Peterson (1958) described a cytoplasmic male sterile system, but this system was unstable and resulted in fertile pollen under cool conditions. Studies using Peterson's CMS material indicate that additional factors affect pollen sterility and stability (Novak *et al.*, 1971; Shifriss and Frankel, 1971; Shifriss and Guri, 1979).

INTERSPECIFIC HYBRIDIZATION

The ability to hybridize between species is important because unique genes from different species may be utilized, e.g. the introgression of tobacco etch virus resistance from *C. chinense* to 'Tabasco' (*C. frutescens*). Interspecific hybridizations between species of *Capsicum* can be made with varying degrees of success (Table 5.1).

BREEDING FOR DISEASE AND PEST RESISTANCE

Peppers are affected by several diseases and pests (see Chapter 10). Cultural methods and pesticides are applied to ensure a healthy and profitable pepper crop. One of the safest and most efficient means of protecting peppers is through the development of disease- and pest-resistant cultivars. Introducing disease resistance into a pepper cultivar also implies that the cultivar meets the horticultural requirements for its market. It is a laborious task to introgress resistance while maintaining horticulturally acceptable characteristics. Therefore, many years may be required before a resistant cultivar can be released, and this task is made even more difficult and time-consuming if the genetic nature of the resistance is quantitative, that is if it is controlled by many genes.

Table 5.1. Interspecific cross-compatibility of *Capsicum* species. (Information from Pickersgill, 1971; Tong and Bosland, 1997.)

Male/female	C. annuum complex					C. baccatum complex			C. pubescens complex		
	C. annuum	*C. chinense*	*C. frutescens*	*C. chacoense*	*C. galapagoense*	*C. baccatum*	*C. praetermissum*	*C. tovarii*	*C. pubescens*	*C. cardenasii*	*C. eximium*
C. annuum complex											
C. annuum		+/+	+/+	+/+	*/+	+/+	*/*	0/0	0/*	0/*	0/*
C. chinense			+/+	+/+	#/#	+/+	+/*	0/#	0/*	NA/NA	0/#
C. frutescens				0/+	0/*	+/+	NA/NA	0/#	NA/#	NA/+	NA/+
C. chacoense					NA/*	*/+	*/NA	NA/*	NA/*	NA/*	NA/*
C. galapagoense						*/NA	NA/NA	NA/*	NA/NA	NA/NA	NA/NA
C. baccatum complex											
C. baccatum							+/+	0/+	NA/*	NA/+	NA/+
C. praetermissum								0/+	NA/*	NA/*	NA/NA
C. tovarii									0/0	0/0	0/0
C. pubescens complex											
C. pubescens										*/+	*/+
C. cardenasii											*/+
C. eximium											

+ represents F$_1$ hybrids that germinate normally.

\# represents F$_1$ hybrids requiring embryo rescue.

* represents F$_1$ hybrids for which inviable seed is produced.

0 represents no fruit and/or seed produced.

/ represents reciprocal cross, i.e. data to left of / represents cross where the male is the species in the left column.

NA, information not available.

Successful cultivars have been developed with resistance to a wide range of pests and pathogens. Currently, cultivars are available that are resistant to nematodes, viruses, fungi and bacteria. Several pepper cultivars have multiple disease resistances. Because resistances to pests and diseases may be overcome by a given pest or pathogen, plant breeders are constantly breeding new pepper cultivars with improved resistance. Sources of resistance that breeders can utilize include established resistant cultivars, landraces, wild relatives and closely related species. Plant breeders prefer to utilize sources of resistance that exist in germplasm that is similar to the type of cultivars being developed. This is simply because it will take less time to produce a cultivar with horticulturally acceptable traits including the desired pest or disease resistance than if an interspecific hybridization were made.

BIOTECHNOLOGY

Biotechnology is being used in pepper to develop new cultivars, map genomes and evaluate genetic resources. At present, genetically engineered pepper is not commercially possible because of the lack of a transformation system. However, there are several areas of biotechnology that are aiding in the development of improved pepper cultivars.

Molecular Markers

Molecular markers have proved to be invaluable for understanding the genetic make-up of agricultural crops. Molecular markers take advantage of technologies that allow scientists and plant breeders to observe genetic differences between two or more individuals. Molecular markers are similar to genetic markers. Genetic markers are seen as morphological differences. Morphological differences have been used since the turn of the 20th century to build genetic maps (Paterson *et al.*, 1991). Molecular markers differ from genetic markers in several ways: (i) molecular markers usually occur in greater numbers; (ii) molecular markers can be distinguished without relying on complete development of the plant; that is, tissue from a plantlet may be analysed rather than waiting for the plant to exhibit some morphological feature; and (iii) the environment does not alter the expression of a molecular marker (Tanksley, 1983). In general, molecular markers are commonly used to examine genetic diversity, systematics and phylogeny. They are used in combination with other markers to construct genetic maps and are used in linkage studies. Markers linked to a desired trait can be used by plant breeders in marker assisted selection (MAS). When markers are identified with a gene or genes of interest, the marker(s) can be used as a selection criterion by plant breeders (Staub *et al.*, 1996). Selection via molecular markers eliminates the need for costly and sometimes inefficient screenings and speeds up the process of cultivar development.

A variety of molecular markers have been used in pepper. Different molecular markers vary in their strengths and weaknesses; therefore, each molecular marker and how it has been used in *Capsicum* research is described below.

Isozymes

Isozymes are protein molecules that are separated electrophoretically based on their charge. Gels are stained for specific enzyme activity and, by doing so, allelic and non-allelic proteins can be identified (Tanksley, 1983.) A major shortcoming of isozyme analysis is the small number of isozyme loci available. The use of isozymes in pepper research has focused predominantly on measuring genetic variability, and clarifying systematic and phylogenetic relationships in the genus.

McLeod *et al.* (1979) used isozymes to examine the systematic relationships of *Capsicum cardenasii*, *Capsicum eximium*, *Capsicum tovarii* and *Capsicum pubescens*. *Capsicum pubescens* is the only domesticated member of the four species in this study. The authors used 230 individual plants from 102 populations across the four species examined. Isozyme analysis yielded 25 loci, coding for 15 proteins. The isozyme data indicated that *C. tovarii* should be classified as a distinct species. They also proposed that *C. cardenasii* and *C. eximium* be grouped together as a single species. Lastly, the authors proposed that *C. pubescens* be maintained as a unique species, but that it be grouped with *C. cardenasii* and *C. eximium* in the same species complex. Analysis of the isozyme data indicated that *C. pubescens* clustered more distantly to *C. cardenasii* and *C. eximium* than those two did to one another, but that *C. pubescens* was more closely related to *C. eximium* than to *C. cardenasii*.

In 1983 McLeod *et al.* expanded their taxonomic treatment via isozyme analysis to a larger segment of the *Capsicum* genus. A total of 1010 individuals from 212 populations were analysed. In this study, 15 enzymatic and non-enzymatic proteins encoded by 26 genetic loci were examined. As a result of isozyme analysis, the authors reported a hypothetical model in which three independent events led to three groups of domesticated peppers. They proposed that *Capsicum baccatum* var. *pendulum* arose from the wild form *C. baccatum* var. *baccatum* in lowland Bolivia. They re-asserted that *C. pubescens* evolved from a *C. eximium/C. cardenasii* ancestral group, and they proposed that the third group, composed of domesticated *C. annuum*, *C. chinense* and *C. frutescens*, originated from a single common wild ancestor in lowland Amazonia.

Conciella *et al.* (1990) analysed the esterase isozymes in 15 accessions of *C. annuum* from the United States, Mexico, Central America and South America. The authors also compared cytological data to the isozyme data. These accessions included both wild and cultivated types of pepper. Analysis of this data showed three different isozyme patterns. All of the Mexican accessions from central Mexico and some of the accessions from the United States shared one pattern. Accessions from Peru, Central America and the other accessions from the United States all shared a second pattern of isozyme banding. The third pattern of isozymes included accessions from Colombia. Interestingly, the third pattern of isozyme bands were identical to those of the hybrids produced when accessions from the first two pattern groups were crossed with one another. The isozyme data combined with the cytological data provide evidence for the theory that Mexico is the centre of domestication for *C. annuum*.

Loaiza-Figueroa *et al.* (1989) used isozymes to study genetic diversity, as well as phylogenetic relationships between accessions from Mexico in more detail. Their research used 186 accessions of *Capsicum* from Mexico. These accessions included *Capsicum annuum* as well as other species and utilized wild, semi-domesticated and domesticated varieties. The nine enzyme systems produced 76 alleles which represented 20 loci, indicating that individual *Capsicum* populations in Mexico are for the most part homozygous, and that isozyme groupings correlated to geographic origin. Comparing similarity indices of semi-domesticated and wild forms, the authors narrowed the theoretical centre of domestication of *C. annuum* to eastern Mexico, in the states of Tamaulipas, Nuevo Leon, San Luis Potosi, Veracruz and Hidalgo.

Isozymes have also been used for genetic mapping. Tanksley (1984b) developed the first isozyme linkage map of *Capsicum* by studying segregating isozymes in a hybrid cross between *C. annuum* and *C. chinense*. Some isozymes were linked to one another and provided additional information for mapping the *Capsicum* genome. Tanksley also compared electrophoretic patterns and intensities of five different trisomic F_1 hybrids with diploid F_1 hybrids in an attempt to detect possible differences in allele dosages and thereby pinpoint chromosomal location of the isozyme encoding genes. Finally, comparison of linkage segments in pepper and tomato (*Lycopersicon*) provided initial information on the similarities and differences between the two genera and how they might have diverged in terms of chromosome evolution.

RFLPs

Restriction fragment length polymorphisms (RFLPs) utilize restriction enzymes that cut genomic DNA at specific sites. The cut DNA fragments are separated by electrophoresis then transferred and immobilized on to nitrocellulose paper. The fragments are then probed, usually with cloned, radioactively labelled probe DNA fragments which are typically 500–3000 base pairs long (Staub *et al.* 1996). RFLPs can distinguish homozygous from heterozygous individuals, but they are expensive, require technical expertise and have the further disadvantage of utilizing radioactive material. In *Capsicum* research, RFLPs have been used for genetic mapping.

Tanksley *et al.* (1988) continued studies into the genomic similarities and differences between *Capsicum* and *Lycopersicon* by using RFLPs to construct the first RFLP linkage map of pepper. The clones used as RFLP probes included cDNA clones derived from tomato leaf RNA, cDNA clones from pepper leaf mRNA, random single-copy tomato genomic clones and clones from other known genes, such as those for the small subunit ribulose biphosphate carboxylase, chlorophyll a/b binding polypeptide, alcohol dehydrogenase and the 45S ribosomal subunit.

A linkage map for pepper was made by looking at the segregation of polymorphic RFLP patterns in 46 individual plants derived from the F_1 generation of a cross between *Capsicum annuum* and *Capsicum chinense* backcrossed to *C. annuum*. All cDNA clones, including clones derived from tomato, hybridized to pepper DNA. Using 80 of the 85 loci detected, 14 linkage groups were constructed for pepper. These results indicate that tomato and pepper generally share the same repertoire of genes. However, the order of the genes on the linkage groups of pepper was discovered to be very different from the order of genes on the tomato linkage map. Comparing the two maps showed that a minimum of 32 breakages of tomato chromosomes would have had to occur to account for the positions of those orthologous genes in the pepper genome.

In 1993, some of the members of the group that developed the first RFLP linkage map developed a more complete linkage map of pepper (Prince *et al.*, 1993). These researchers used three times the number of markers used in their preliminary map. They used random genomic clones from tomato in addition to cDNA clones as probes for RFLP analysis. They used 46 plants derived from selfing the F_1 hybrid between *Capsicum annuum* and *Capsicum chinense*. The map utilized 192 markers in 19 linkage groups with a total coverage of 720 cM (centiMorgans). However, specific map positions of 26 RFLP markers in seven linkage groups were not determined and vast regions of the pepper genome remained unmapped. Two quantitative trait loci were also mapped and more detailed comparisons between tomato and pepper were made.

In 1993, Lefebvre *et al.* demonstrated that RFLPs were more useful than isozymes for mapping and diversity studies within a *Capsicum* species. These researchers used RFLPs to look at genetic variability within *C. annuum* var. *annuum*. They analysed 13 inbred lines from landraces that were collected from a wide geographical range, inbred bell pepper varieties and an accession of *C. baccatum* that served as an outlier species. Statistical analysis of the data indicated that enough variability existed within *C. annuum* to construct an intraspecific map. The researchers intended to use this data to build a map of the species and interpret the results of their recurrent selection programme for resistance to diseases in pepper.

RFLPs have also been used to study genetic diversity. In 1992, Prince *et al.* expanded the isozyme studies of Loaiza-Figueroa *et al.* (1989). RFLP analysis was performed on 25 accessions from Mexico and two parents were used for molecular mapping. The results of the study confirmed those of Loaiza-Figueroa *et al.* The analysis separated the Mexican accessions into two main groups based generally on species and geographic origin. The southern Mexican accessions, composed of *C. frutescens*, formed a distinctly different group from those of northern Mexico, which were composed of *C. annuum* accessions. Based on the results of the experiment, the authors also selected accessions that would be useful for RFLP mapping efforts within *C. annuum*.

In 1994, Prince *et al.* expanded their survey of interspecific genetic variation using RFLPs. The authors used 21 diverse accessions of cultivated and wild pepper, representing five genera (*C. annuum, C. baccatum, C. chacoense, C. chinense*, and *C. frutescens*). The RFLP analysis differentiated each accession in the study and demonstrated that any two accessions could be used as possible parents for RFLP mapping. The authors combined RFLP data with data from a relatively new technique called random amplified polymorphic DNA (RAPD). The authors used these data to evaluate intraspecific variablity between four *C. annuum* cultivars.

RAPDs

Randomly amplified polymorphic DNA (RAPD) utilizes the polymerase chain reaction (PCR). Polymorphic markers are generated using single primers which are usually ten base pairs long (Williams *et al.*, 1990). RAPDs have been used in *Capsicum* research to study genetic diversity, linkage and to provide additional molecular markers for mapping.

As described above, Prince *et al.* (1994) utilized RAPD molecular markers as well as RFLPs in their study of molecular marker polymorphism within *C. annuum*. The two molecular marker techniques were also compared with one another in terms of use in DNA fingerprinting and differentiation of genetic distances. The authors concluded that both techniques are useful in understanding genetic diversity at the cultivar level within a species such as *C. annuum*. RAPDs were also deemed useful for DNA fingerprinting and identifying closely related cultivars.

The use of RAPDs in linkage studies was described in 1993 by Inai *et al.* Severe dwarfism was seen in a cross between *C. chinense* and *C. annuum* when *C. chinense* was the female parent. In the reciprocal cross, 10 of 16 showed dwarfism, leading to the assumption that the dwarf phenotype occurred as a result of an interaction between *C. annuum* nuclear genes and *C. chinense* cytoplasm. In order to evaluate the nuclear component, the authors screened and found a linked RAPD molecular marker that had a recombination frequency of 6%.

In 1997, Lefebvre *et al.* used RAPDs, RFLPs, known function genes, an isozyme and phenotypic markers to develop an intraspecific molecular linkage map developed from the F_1 hybrid derived from two double haploid *C. annuum* populations.

GENETIC ENGINEERING

Genetic engineering in pepper is dependent upon a reliable means of transformation and tissue culture regeneration. At present, research in the area of genetically transforming pepper has been limited by the lack of a reproducible, efficient and reliable means of regenerating plants via tissue culture. Once a standardized protocol for tissue culture regeneration of pepper plantlets is developed, researchers will probably be able to insert desired segments of foreign DNA into *Capsicum* genomes via *Agrobacterium tumefaciens*-mediated transformation. Preliminary research into genetically transforming *Capsicum* has resulted in a few transformed sweet and hot pepper plantlets. Research by Zhu *et al.* (1996) has produced cucumber mosaic virus resistance in chile plantlets by integrating genes responsible for various protein components of the virus itself into the targeted plant's genome.

TISSUE CULTURE

Tissue culture of pepper can itself be divided into different areas of research: anther culture, protoplast regeneration, embryo rescue and organogenesis. Anther culture involves regenerating plantlets from pollen microspore tissue. Plantlets produced from pollen microspores are haploid. Using colchicine, the chromosome number can be doubled. The benefit derived from double haploid plants, as they are called, is that these plants are homozygous at every loci. Many generations of self-pollination would be required to produce this same effect. Plants that are homozygous at all loci are invaluable for a variety of research needs.

Research on anther culture in pepper started in the early 1970s and, to date, it is the most widely tested system of tissue culture in *Capsicum*. In 1973, John-Shang *et al.* reported that red pepper anthers could be cultured on standard media and growth substances. The anthers produced embryoids and calli, and a few seedlings were generated. Cytological examination of the root tips revealed that some of the seedlings were, in fact, haploid in nature.

In 1974, Harn *et al.* were able to induce callus and embryoids from anthers of 'Kimjang Kochu', a Korean hot pepper variety. The anthers were cultured on Murashige and Skoog media, and a variety of growth regulators were tested. Somatic calluses were formed from the connective, filament and inner tissues of the anther. Haploid calluses and embryoids were produced from microspore tissue within the anther locule, especially when the anthers were at the late uninucleate stage. Researchers in France reported that one to three plantlets could be produced from every 100 cultured anthers (Sibi *et al.*, 1979). The process was unique in that it involved a cold pretreatment of buds at 4°C for 48 h and a transfer of anthers after 12 days of culture to a different media. The chromosome numbers of 24 plants were as follows: 20 were haploid, two were diploid and two were triploid.

An improved method that produced higher numbers of plants from anther culture was described in 1981 by De Vaulx *et al.* Increasing success was attributed to a 35°C temperature in darkness for the first days of culture and a transfer of anthers after 12 days to a new media. From 5 to 40 plants per 100 cultured anthers were produced and high frequencies of haploid and haploid/diploid chimera were observed. In 1995, Nervo *et al.* investigated the homogeneity and genetic stability of double haploid lines produced from anther culture. Morphological and molecular marker characterization were made on double haploid pepper plants. This research demonstrated that double haploid plants are homozygous and genetically stable after selfing.

Protoplast Regeneration

Protoplast regeneration involves isolating single plant cells whose cell walls are digested away. Protoplasts may be fused with one another to form cybrids, or they may be subject to other techniques in which transformation is intended. In 1981, Saxena *et al.* isolated protoplasts from mesophyll cells of the cultivar 'California Wonder'. Mitotic division and callus formation were observed followed by differentiation into whole flowering plants. Diaz *et al.* (1988) reported isolation of protoplasts from four genotypes of *Capsicum annuum* and one of *Capsicum chinense*. Whole plants were reported as being regenerated in one cultivar, 'Dulca Italiano'.

Embryo Culture

Embryo culture is useful when 'embryo rescue' is required. Embryos from interspecific hybridizations often abort before seed development is complete. Some interspecific hybridizations may produce viable plants if the embryo is rescued at an early stage of development. This technique of excising the embryos and placing them on a nutrient medium has been accomplished in *Capsicum* (Fari, 1995).

Organogenesis

Organogenesis involves removing plant tissue such as hypocotyls or cotyledons from germinating seeds and placing these explants on tissue culture media in order to induce differentiation and development of organs and plantlets. Gunay and Rao (1978) are credited with the first successful regeneration of plantlets via organogenesis. The authors used three varieties of *Capsicum* from which they excised cotyledon and hypocotyl explants. Shoot buds and roots were regenerated under specific hormone treatments.

In 1981, Fari and Czako examined different sections of hypocotyl to determine suitability for organogenesis. They determined that different segments responded differently in culture. Apical sections produced shoot buds only, while middle sections formed roots, and basal sections produced callus. Phillips and Hubstenberger (1985) further examined culture parameters suitable for organogenesis in pepper cultivars (Figs 5.1–5.4). The authors concluded that light and temperature were important factors in shoot and root organogenesis. They also found that low levels of certain hormones were required for shoot elongation and rooting, yet high levels of one hormone were necessary for adventitious bud formation. Finally the authors demonstrated that glucose was preferable over sucrose as a carbon source.

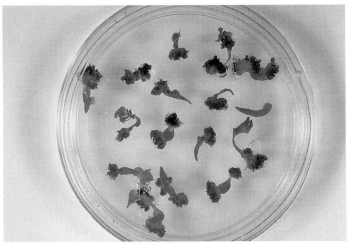

Fig. 5.1. Cotyledon explants with callus and differentiating tissue (Phillips and Hubstenberger, 1985).

Fig. 5.2. Later development of callus and differentiating tissue (Phillips and Hubstenberger, 1985).

CONCLUSION

A peculiar aspect of peppers is their inability to be regenerated from protoplasts. This limits the technique of genetic transformation. Within other solanaceous crops, e.g. tomato, tobacco and petunia, excellent progress has been made

Fig. 5.3.   Explant with stem, leaves and roots (Phillips and Hubstenberger, 1985).

Fig. 5.4.   Plants in greenhouse produced from explants (Phillips and Hubstenberger, 1985).

regenerating plants and using genetic transformation to introduce novel genes into the genome. For unexplained reasons, pepper has been recalcitrant to regeneration. Many laboratories around the globe are addressing this research question. Once this perplexing problem is solved, genetic transformation will be available for the introduction of novel genes in peppers. Classical methods, consisting of direct selection within a segregating population of resistant pepper plants and inoculation at the seedling stage under appropriate environmental conditions, will continue to be the main thrust for incorporating disease and pest tolerance. Indirect selection based on linked molecular markers offers new opportunities. The use of RAPD markers along with chromosome pairing behaviour in interspecific crosses is being used to establish the interrelationships among the *Capsicum* species. The increased demand for new pepper cultivars will require the sharing of ideas and germplasm allowing the pepper industry to grow and prosper.

6

CHEMICAL COMPOSITION

As with most plants, peppers contain thousands of chemicals including water, fixed (fatty) oils, steam-volatile oil, carotenoids, capsaicinoids, resin, protein, fibre and mineral elements. The numerous chemicals have importance for nutritional value, taste, colour and aroma. The two most important groups of chemicals found in peppers may be the carotenoids and the capsaicinoids. The carotenoids contribute to a pepper's colour and its nutritional value. The capsaicinoids are the alkaloids that give hot peppers their characteristic pungency.

WATER

Water is the most plentiful chemical in peppers. In pepper fruits, the amount of water is dependent on the age and type of pod harvested (see Table 6.1). Green mature pods contain the largest amount, about 90%, while spice varieties allowed to dry on the plant may contain as little as 70%. A dehydrated product prepared for storage or shipping is reduced to 15–20% water. Peppers rapidly lose water after harvest, contributing to a major quality problem when

Table 6.1. Physical attributes of three pepper pod types. (Data from Lownds *et al.* 1993.)

Pod type	Initial water content (%)	Surface area (cm^2)	Surface volume ratio	Cuticle weight (mg cm^{-2})	Epicuticular wax (µg cm^{-2})
Bell	92.1a^z	553a	0.88c	1.8c	113a
New Mexican	90.6b	340b	1.78b	4.4a	76.4b
Yellow wax	92.0a	270c	2.77a	2.4b	55.5b

^{z}Means within a column followed by different letters are significantly different at $P \le 0.01$ with LSD procedure.

shipping fresh pepper pods (Lownds *et al.*, 1993). New Mexican-type peppers lose water twice as fast as bell or jalapeño types (Lownds *et al.*, 1994).

CARBOHYDRATES

Pepper fruits contain sugar, pentosans and raw fibre. Glucose accounts for 90–98% of the sugar content of red mature paprika pods (Somos, 1984). The amount of sugar in pods varies by cultivar and pod type. Some add an appreciable 'sweetness' to the fruits, while other types completely lack the sensation of sweetness. Total and reducing sugars are at maximum levels in red succulent fruits (Wall and Biles, 1993).

Cellulose and other fibrous material may account for up to 20% of the dry weight of pericarp tissue. A by-product of the canning of green New Mexican pods is the pod skins. McKee (1998a) found that the skins contained 77% soluble fibre and 80% total dietary fibre. This amount of fibre is greater than in either rice or oats. She concluded that green New Mexican skins could be an excellent source of dietary fibre if added to processed baked goods.

LIPIDS

Peppers contain lipids that are qualitatively similar to lipids found in plants in general. The total lipid content of fresh, green, bell pepper pods is 400 mg 100 g^{-1} (wet wt), a relatively low level (Kinsella, 1971). When the lipids were examined for composition, they were found to contain 82% neutral lipids (fats), 2% phospholipids and 16% glycolipids. The triglycerides made up 60% of the total lipids with palmitic acid, linoleic acid and linolenic acid being preponderant (Lyons and Lippert, 1966). The phospholipids were 76% phosphatidylcholine. The linoleic acid accounted for 70% of the fatty acid composition.

In addition, the amount of unsaturated fatty acids in the mitochondria relates to the sensitivity of a species to chilling injury. Plants with higher amounts of unsaturated fatty acids are more resistant to chilling temperatures than those with more saturated fatty acids. Peppers are sensitive to chilling temperatures and the ratios of unsaturated/saturated fatty acids are those expected for a chilling-sensitive species.

AMINO ACIDS, PROTEINS, MICROELEMENTS

Somos (1984) details several studies undertaken in Hungary on this group of chemical compounds. He lists lysine, arginine, proline, tyrosine, tryptophan, methionine, valine, phenylalanine, leucine, glutamic acid, glycine, asparagine, threonine and alanine as being found in pepper fruits. He also states that research in Hungary has shown that the pericarp has 16–17% protein and the seeds contain 18% protein. When the microelements were investigated, it was found that iron was present in the largest concentration, followed by bromide and manganese. Other microelements found were cadmium, calcium, cobalt, copper, magnesium, phosphorus, potassium, sodium and zinc.

CAPSIDIOL

Much research has been published on one of the phytoalexins found in pepper, capsidiol. The chemical is a bicyclic sesquiterpene which is synthesized through the mevalonic acid pathway. Capsidiol was shown to be a fungistatic chemical to the pathogen, *Phytophthora capsici* (Jones *et al.*, 1975). The importance of this post-infectional chemical is debated. Capsidiol and its direct relation to disease resistance has not been elucidated.

FLAVOUR/AROMA

Most peppers are used for flavour, not heat. Flavour is a complex sensation determined in the mouth. Pepper connoisseurs can readily identify subtle flavours presented by each type. As in wine-tasting, one can distinguish between the subtle flavours of peppers after a few years of experience: ancho is sweetish, mulato is chocolaty, mirasol is fruity and chipotle is smoky. One of the most potent volatiles known to humankind is found in pepper, the pyrazine 2-methoxy-3-isobutyl-pyrazine, the 'green bell pepper' smell. Buttery *et al.* (1969) found that humans can detect this odour at 2 parts per trillion. Keller *et al.* (1981) surveyed pepper volatiles and found 102 odour compounds in *C. annuum* and *C. frutescens*. In a similar study, Haymon and Aurand (1971) found that the oil extracted from *C. frutescens* cv. 'Tabasco' contained 125 components whose relative abundance changed with the season of harvest. The composition of aroma compounds of tabasco differed significantly from that of green bell pepper. The tabasco sample contained no pyrazine compounds. To reconstitute the tabasco aroma it took three main chemicals, 4-methyl-1-pentyl-2-methylbutyrate, 3-methyl-1-pentyl-3-methylbutyrate, and isohexyl-isocaproate.

VITAMINS

Peppers are good sources of several vitamins. They produce high concentrations of vitamin C, provitamin A, E, P (citrin), thiamine (B_1), riboflavin (B_2), and niacin (B_3). A wide range of vitamin levels have been reported and this phenomenon has been attributed to differences in cultivars, maturity, growing

Fig. 6.1.   Red mature New Mexican pods on a plant in the field.

practices, climates, postharvest handling and analytical methods (Mozafar, 1994). Provitamin A is discussed under the topic of carotenoids. For thiamine, riboflavin and niacin, reported amounts of each vary depending upon pod type examined. Thiamine amounts range from 0.60 to 0.40 mg 100 g^{-1}. Riboflavin had a range of 0.93–1.66 100 g^{-1}, and niacin amounts were 13.6–15.4 mg 100 g^{-1} (Govindarajan, 1988).

Pepper is an extremely rich source of ascorbic acid (vitamin C); one of the richest sources of vitamin C in vegetables. The ascorbic acid content has been reported at between 46 and 243 mg 100 g^{-1} fresh weight (Wimalasiri and Wills, 1983; Nisperos-Carriedos *et al.*, 1992; Howard *et al.*, 1994; Lee *et al.*, 1995) (see Fig. 6.2). Ascorbic acid increases during fruit ripening (Osuna-Garcia, 1996). Levels peak at different maturity stages depending on the cultivar. The differences among cultivars can be attributed to variation in moisture content of the fruits. Ascorbic acid is a water-soluble compound that can be expected to decline as the fruits dehydrate. Fruit from the green to the red succulent stage (Fig. 6.1) all contain enough ascorbic acid to meet or exceed the adult RDA (60 mg) for vitamin C (NRC, 1989).

Peppers are rich sources of vitamin E with the tocopherols being the source of vitamin E. Red, dry pepper powder has α-tocopherol levels comparable with those for spinach and asparagus, and on a dry weight basis has four times more than tomatoes. On a dry weight basis, a 100 g red fruit would exceed the RDA (8–10 mg) for the average adult (NRC, 1989). They range in concentrations from 3.7 to 236 mg 100 g^{-1} dry wt depending on the reference source (Kanner *et al.*, 1979; Daood *et al.*, 1989; Biacs *et al.*, 1992). Osuna-Garcia (1996) reported that pepper seeds contained γ-tocopherol, while the pericarp contained α-tocopherol. He also reported that the γ-tocopherol reached its

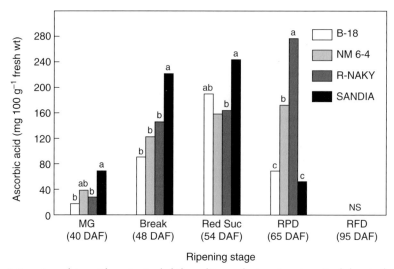

Fig. 6.2. Ascorbic acid content of chile cultivars during ripening. Each bar is the mean of five replications. Cultivar means within ripening stages having the same letter are not significantly different ($P > 0.05$). MG, mature green; break, breaker; Red Suc, red succulent; RPD, red partially dry; RFD, red fully dry; DAF, days after flowering (Osuna-Garcia *et al.*, 1998).

maximum in seeds of fruit at the red succulent stage (41.7 mg 100 g^{-1} dry wt), and then declined. Kanner *et al.* (1979) reported that the content of α-tocopherol is dependent on the lipid content, which varies according to ripening stage and cultivar. Kanner *et al.* (1979) and Osuna-Garcia (1996) reported that α-tocopherol increased during ripening from the mature green (3.9 mg 100 g^{-1} dry wt) to red fully dry stages (23.8 mg 100 g^{-1} dry wt). Osuna-Garcia (1996) also found that cultivars differed in the amount of each tocopherol with some cultivars having significantly higher levels (see Figs 6.3 and 6.4).

CAROTENOIDS

The diverse and brilliant colours of pepper fruit originate from the carotenoid pigments present in the thylakoid membranes of the chromoplasts (see Table 6.2). In plants, carotenoids are synthesized in both the chloroplasts of photosynthetic tissues and the chromoplasts of flowers, fruit and roots. Chemically, carotenoids are lipid-soluble, symmetrical hydrocarbons with a series of conjugated double bonds. The double bond structure is responsible for the absorption of visible light. Carotenoids function as accessory pigments for photosynthesis but, more importantly, as photoprotectants in the plant. The primary function of β-carotene and other carotenoids is to protect the

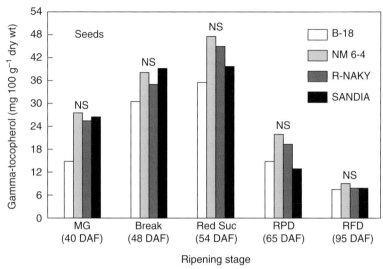

Fig. 6.3. γ-Tocopherol content of chile cultivars during ripening. Each bar is the mean of five replications. Cultivar means within ripening stages are not significantly different ($P > 0.05$). MG, mature green; break, breaker; Red Suc, red succulent; RPD, red partially dry; RFD, red fully dry; DAF, days after flowering (Osuna-Garcia *et al.*, 1998).

chloroplasts from photo-oxidative damage. However, carotenoids are unstable when exposed to light, oxygen or high temperatures. The carotenoids in the fruit are important for attracting seed dispersers (birds).

The green, yellow, orange and red colours originate from the carotenoid pigments produced in the fruit during ripening. More than 30 different pigments have been indentified in pepper fruits (Matus *et al.*, 1991). These pigments include the green chlorophylls (a and b); the yellow-orange lutein, zeaxanthin, violaxanthin, antheraxanthin, β-cryptoxanthin and β-carotene; and the red pigments, capsanthin, capsorubin and cryptocapsin, which are found only in pepper fruits. The red colour in pepper comes from the carotenoids capsanthin and capsorubin, while the yellow-orange colour is from β-carotene and violaxanthin. Capsanthin, the major carotenoid in ripe fruits, contributes up to 60% of the total carotenoids. Capsanthin and capsorubin increase proportionally with advanced stages of ripeness, with capsanthin being the more stable of the two (Harkay-Vinkler, 1974; Kanner *et al.*, 1979). The amount of carotenoids in fruit tissue depends on factors such as cultivar, maturity stage and growing conditions (Reeves, 1987).

In peppers, 95% of the total provitamin A in green pods and 93% in mature red pods is β-carotene (Howard *et al.*, 1994). When red mature pods were measured, the cultivars with the highest and the lowest provitamin A activity were both yellow wax pod types. Howard *et al.* (1994) measured the mature red pods

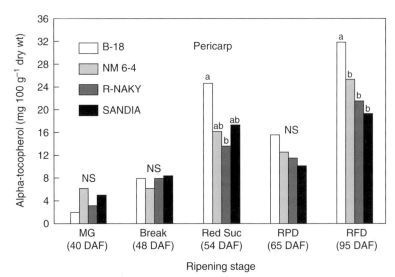

Fig. 6.4. α-Tocopherol content of chile cultivars during ripening. Each bar is the mean of five replications. Cultivar means within ripening stages having the same letter are not significantly different ($P > 0.05$). MG, mature green; break, breaker; Red Suc, red succulent; RPD, red partially dry; RFD, red fully dry; DAF, days after flowering (Osuna-Garcia *et al.*, 1998).

Table 6.2. The most common carotenoid pigments found in peppers.

Yellow food colouring agents
   Antheraxanthin
   β-carotene
   β-cryptoxanthin
   Lutein
   Violaxanthin
   Zeaxanthin
Orange-red food colouring agents
   Capsanthin
   Capsorubin
   Cryptocapsin

of the yellow wax pod type and found the α-, β-carotene and provitamin A activity increased by 344%, 255% and 229%, respectively, as the pods matured.

ANTI-OXIDATIVE PROPERTIES

Capsicum does have a strong anti-oxidative property, and the binding of free radicals may be the mechanism (Colditz, 1987). Several mechanisms for the possible protective action of β-carotene have been suggested (Peto, 1983). More than 20 carotenoids have been isolated from peppers. Recent evidence for antitumour activity of carotenoids in humans has renewed interest in foods containing these compounds, especially because diet supplements do not provide the same nutritional and medicinal benefits as fresh fruits and vegetables. The mode of action for the carotenoids appears to be that they are radical scavengers, effectively binding the singlet or excited oxygen and the free radicals, which may cause damage in humans under physiological conditions of oxygen tension (Burton and Ingold, 1984), a characteristic not shared by retinol.

PUNGENCY

Besides colour (carotenoids), another important quality attribute of pepper is pungency (heat). Some have argued that pungency is one of the five main taste senses, along with bitter, sweet, sour and salty. Physiologically, the senses responsible for our perception of flavour can be divided into three anatomical systems. In the oral cavity, the classical gustatory pathways through the tongue and soft palate are responsible for our sensitivity to the four basic tastes, sweet, sour, salty and bitter. In the nasal passages, the olfactory receptors provide sensitivity to a wide variety of volatile compounds, producing the sensations we normally assign to smell. In addition to these two systems, the trigeminal nerves in both the oral and the nasal cavity provide sensitivity to thermal, tactile, irritation and pain sensations. The trigeminal innervation is also chemically sensitive to compounds that are pungent, and hence provide an important part of our appreciation of flavour as a whole. Because the capsaicinoids are potent stimuli of the oral trigeminal nerves, they are a desirable attribute of many foods. In most parts of the world, pungency makes otherwise bland staple foods appetizing.

The word 'pungency' can be confusing. Some prefer the terms 'hot flavour', heat, fiery or spicy to pungency. In this book, the sensory response is identified as pungency and the substance responsible for pungency by its chemical name, capsaicin, dihydrocapsaicin, etc. The capsaicinoids can be analysed or estimated by their physical or chemical characteristics, but the pungency of a pepper product can only be validated through a correlation with the perceived heat associated with oral consumption. This has become most relevant in recent years because of the complexity of the food items containing capsaicinoids.

The nature of the pungency has been established as a mixture of seven or more homologous branded-chain alkyl vanillylamides, named capsaicinoids (Torabi, 1997). The capsaicinoids are unique to the *Capsicum* genus (see Table 6.3). They are odourless, colourless, flavourless, non-nutrient compounds. Capsaicin composition ($C_{18}H_{27}NO_3$) is similar to peperin ($C_{17}H_{19}NO_3$), which gives black pepper its bite. The capsaicinoids are produced in glands on the placenta of the fruit (see Fig. 6.5). While seeds are not the source of the pungency, they occasionally absorb capsaicin because of their proximity to the placenta.

The capsaicinoids are not sensed by the taste buds. Heat sensation from the capsaicinoids results from irritation of the trigeminal cells, which are pain

Table 6.3. The known capsaicinoids in *Capsicum*. Sources for the identification of the capsaicinoids: [a]Bennett and Kirby (1968), [b]Collins and Bosland (1994), [c]Collins *et al.* (1995), [d]Torabi (1997), [e]Krajewska and Powers (1988), [f]Kaga *et al.* (1992).

Capsaicin[a]	Dihydrocapsaicin[a]
Homocapsaicin[a]	Isomer of dihydrocapsaicin[c]
Homocapsaicin II[d,e]	Homodihydrocapsaicin[a]
Norcapsaicin[f]	Homodihydrocapsaicin II[d,e]
Nornorcapsaicin[f]	3-nor-dihydrocapsaicin[d]
Bis-homocapsaicin[d e]	Nordihydrocapsaicin[a]
Tris-homocapsaicin[d,e]	Nornordihydrocapsaicin[b]
Tetra-homocapsaicin[b]	Isomer of nordihydrocapsaicin[d]
Tetra-homodihydrocapsaicin[b]	Isomer of nornordihydrocapsaicin[d]
Isomer of tetra-homodihydrocapsaicin[d]	Bis-homodihydrocapsaicin[d]
Tris-homodihydrocapsaicin[b]	Isomer of tris-homodihydrocapsaicin[d]

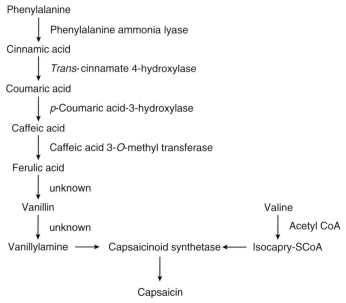

Fig. 6.5. Proposed biosynthetic pathway for the capsaicinoid, capsaicin (Zewdie-Tarekegn, 1999).

receptors located in the mouth, nose and stomach (Silver and Maruniak, 1981). The capsaicinoids stimulate and then desensitize specific subpopulations of sensory receptors (Szallasi and Blumberg, 1990). The receptors release a chemical messenger, substance P, which signals the brain about pain. Even at dilutions down to 1 part per 16 million a sensation of warmth can be detected (Jurenitsch, 1981). The nervous system telegraphs a signal to the brain to flood the nerve endings with endorphins, which are the body's natural painkillers. Endorphins may be viewed as a natural and safe morphine. It is also the endorphins that cause the effect known as 'runners-high', the good feeling that comes after about 8 km of running. The release of endorphins give the body a sense of pleasure. It has been suggested that the release of endorphins is why people become 'addicted' to peppers. Experiments have been conducted in which an endorphin-blocking drug is injected into the subject after the heat sensation had subsided and the subject again feels burning in the mouth. Thus, the capsaicinoids are not destroyed in the mouth, the body masks their presence.

It has been shown organoleptically that humans not only note intensity of pungency, but perceive each capsaicinoid differently (Krajewska and Powers, 1988). The investigations of Krajewska and Powers (1988) revealed that nordihydrocapsaicin (NDC) was the least irritating, and the burning was located in the front of the mouth and palate. It caused a 'mellow warming effect'. The heat sensation developed immediately after swallowing and receded rapidly. In comparison, capsaicin and dihydrocapsaicin were more irritating and were described as having a 'typical' heat sensation. Both compounds produced the heat in the mid-mouth and mid-palate as well as the throat and the back of the tongue. In contrast, homodihydrocapsaicin was very irritating, harsh and very sharp. The heat did not develop immediately and it affected the throat, back of the tongue and the palate for a prolonged period. The heat sensation can last for up to 6 h after ingestion (personal observation). Different combinations of these capsaicinoids produce the different pungency characteristics of individual pepper varieties.

There are more that 200 papers published on the determination and estimation of capsaicinoids in *Capsicum*, oleoresin and products containing their extracts. Analytical methods for determining pepper colour and pungency are not described in detail here. For a detailed description of analytical methods the reader is referred to Wall and Bosland (1998). The methods can be categorized into five basic groups: (i) organoleptic; (ii) colorimetric methods: chromogenic reagents reacted directly with the phenolic hydroxyl of the vanillyl moiety on the extracts of the fruits; (iii) thin-layer chromatography (TLC) and paper chromatography; (iv) gas chromatography; and (v) high-performance liquid chromatography (HPLC).

The two most common methods to measure pungency are the Scoville organoleptic test and the HPLC (ASTA, 1985; Collins *et al.*, 1995). Pepper pungency is expressed in Scoville heat units (SHU) (Scoville, 1912). The Scoville organoleptic test was the first reliable measurement of the pungency of peppers. This test used a panel of five human subjects who tasted a pepper sample and then recorded the heat level. A sample was diluted until pungency could no longer be detected. The organoleptic method or taste test has been the standard method for pungency analysis. Although this method is widely used, it has limitations. Tasters must be trained and their ability to test many samples is restricted by the heat of the test solution. Taster fatigue is a real phenomenon and tasters are also not able to distinguish between the different capsaicinoids. Therefore, the Scoville organoleptic test has been replaced with instrumental methods.

The most common instrumental method is high-performance liquid chromatography (HPLC). It provides accurate and efficient analysis of the content and type of capsaicinoids present in a pepper sample (Collins *et al.*, 1995). HPLC analysis has become the standard method for routine analysis by the processing industry. The method is rapid and can handle a larger number of samples. A common practice today is to multiply capsaicinoid ppm by 15 to convert to SHU.

It is important to measure capsaicinoid content because of its use in a wide range of industries. The pharmaceutical industry uses capsaicin as a counter-irritant balm for external application (Carmichael, 1991). It is the active ingredient in Heet® and Sloan's Liniment®, two rubdown liniments used for sore muscles. Capsaicin has been prescribed for severe chronic pain conditions where it is usually administered topically for periods of several weeks, as in the case of *Herpes zoster* or 'shingles'. When applied topically to treat skin pain like that from shingles, capsaicin depletes levels of substance P. The depletion action is what causes the 'warming' feeling on the skin. Capsaicin also prevents the nerve endings from making more substance P, thus further pain is diminished or completely eliminated. It has also been shown to be effective against cluster headaches (Sicuteri *et al.*, 1990) and useful for the temporary relief of pain associated with rheumatoid arthritis and osteoarthritis (Deal *et al.*, 1991). One of the minor capsaicinoids, dihydrocapsaicin, protects against serum hyperlipidaemia in guinea pigs fed a cholesterol-enriched diet (Negulesco *et al.*, 1989).

Personal defence aerosols with capsaicin as the active ingredient have replaced mace and tear gas in more than a thousand police departments in the United States. The spray will cause attackers to gasp and be blinded for at least 20 min. Many innovative uses of pungency are being studied. Besides new medicinal applications, it has been tried as a barnacle repellent, to repel mice from gnawing on underground electrical cables and to keep squirrels from eating bird seed.

The pungency level of peppers is determined by genetic and environmental components. The capsaicinoid content is affected by the genetic make-up of the cultivar, weather conditions, growing conditions and fruit age. Plant breeders can selectively develop cultivars within certain ranges of pungency.

Also, growers can, to some extent, control pungency by the amount of stress to which they subject their plants. Pungency is increased with increased environmental stress (Harvell and Bosland, 1997). More specifically, any stress to the pepper plant will increase the amount of capsaicinoid level in the pods. A few hot days can increase the capsaicinoid content significantly. In New Mexico, it has been observed that even after a furrow irrigation, the heat level will increase in the pods. Anthropopathically, the plant has sensed the flooding of its root zone as a stress, and has increased the capsaicinoid level in its pods. If the same cultivar was grown in both a hot semi-arid region and a cool coastal region, the fruit harvested from the hot semi-arid region would be higher in capsaicinoids than the fruits harvested in the cool coastal climate.

So why have peppers evolved pungency? In an evolutionary sense, it may be to encourage the correct animal to eat them and to discourage the wrong ones. Rabbits and other small mammals have digestive tracts that destroy the seed. Most solanaceous plants have alkaloids in sufficiently high levels to be toxic to many mammals. Oddly, peppers do not contain alkaloids in their leaves. In the Philippines, pepper leaves are eaten as a leafy vegetable. Without these alkaloids, pepper may have evolved another strategy for partial protection from the wrong animals and to attract to the right ones. Capsaicin is not toxic *per se*, but it may be ferocious enough when it hits the mouth to discourage mammals. Birds, on the other hand, are attracted to red fruit, and have digestive tracts that chemically and physically soften the seed-coats in the fruits without damaging the seeds, thus encouraging germination. In fact, some seeds will suffer retarded germination if they do not pass through a bird's digestion system. Wild pepper piquins are so strongly associated with birds that a common name for them is bird pepper.

INFLUENCING CHEMICAL COMPOSITION

Traditionally growers have been concerned with those characteristics that relate to yield, particularly disease resistance. Processors are most concerned with those characteristics that deal with processing quality or pack-out[1]. Nutritional quality or chemical composition has not been a principal breeding objective nor an area where much physiology research has been done. The exception with peppers is pungency, where environmental and genetic factors have been extensively studied so as to manipulate the pungency to an acceptable level. This phenomenon of lack of research is not only limited to peppers; it is true for most fruits and vegetables. This is because it is generally believed that the consumer buys fruits and vegetables based on appearance and flavour, not

---

1   Pack-out is the quantity of finished processed product that is made from fresh raw product. For example, pack out of green chile is the amount of green chile material canned, after the skin, calyx, stem and seeds are removed.

nutritional quality. This may be changing as consumers begin to look to fruits and vegetables as insurance against illness. The use of carotenoids as anti-oxidants, or the consumption of green fresh peppers for vitamin C to prevent flu and colds illustrates this trend. It can be expected that future research by geneticists, biotechnologists and plant physiologists will concentrate on this area. In the future, peppers may be genetically transformed to be the carrier agent for vaccines, antibiotics and other therapeutic chemicals.

7

PRODUCTION

INTRODUCTION

Peppers are grown on all the continents except Antarctica. However, even in Antarctica there are stories of researchers who have kept peppers in flower pots to spice up their food. Pepper production is found from the humid tropics, to the dry deserts, to the cool, temperate climates. Pepper can be grown as an annual or a perennial crop. It can be grown outside in fields (see Fig. 7.1) or under protective covers as in greenhouses. The ability of pepper to grow and produce a quality crop in such a wide range of climates has made it a common crop world-wide. Because of pepper's extensive adaptation, it is impossible to list a single production method for pepper production worldwide.

Even with a multitude of different methods, one can generalize production as occurring either in the field or in a greenhouse. The majority of commercial acreage in the world is grown with the use of an integrated pest management (IPM) approach. This approach uses manufactured chemicals to produce the pepper crop, but uses them judiciously, limiting the impact on the environment. When growing the pepper crop in a field, the grower has the option of choosing from direct seeding versus transplanting, rainfall versus irrigation, hand harvesting versus machine. In this chapter, an overview of the most common production methods for pepper is presented.

CLIMATE REQUIREMENTS

Peppers are a warm season crop that require about the same growing conditions as tomatoes and eggplants. It does best with a long, frost-free season to produce good quality, high yields. Pepper is highly susceptible to frost and grows poorly in the 5–15°C temperature range. The optimum temperature for pepper growth and development is higher than that for tomato. Rapid germination and emergence is important to ensure a good stand and adequate

Fig. 7.1. A field of pepper plants in New Mexico, USA.

yields. Pepper seeds germinate slowly if at all in cold soils, but emergence accelerates in soils of 24–30°C. If seeds are planted too early in the season when soil temperatures are too cool, germination rate is slowed, affecting emergence and growth of the seedlings. Slow growth can prolong seedling exposure to insects, diseases, salt or soil crusting, any of which can kill all the seedlings. Higher yields result when daily air temperature ranges between 18–32°C during fruit set. The base growing-degree-days temperature is 18°C and temperatures below 18°C provided negligible growth in pepper plants (Sanders *et al.*, 1980).

Several chemical products have been marketed as inexpensive and effective in preventing crop damage from frost or freeze. Perry *et al.* (1992) evaluated two commercially available materials, FrostFree™ and VaproGard™, for frost and freeze protection of pepper under field conditions. Protection was not observed when minimum air temperature reached –3.5°C and –1.0°C on separate occasions. Neither cryoprotectant injured the foliage in the absence of cold events.

SOIL TYPE

As with most crops, the ideal soil for producing peppers is one described as a deep, well-drained, medium-textured sandy loam or loam soil that holds moisture and has some organic matter. Most peppers are grown in soils with a pH range of 7.0–8.5.

When preparing the land for planting, a soil test is important for maximum pepper production. A soil test determines nitrogen, phosphorus, micronutrient needs, pH, salts, electrical conductivity and organic matter content of the soil. A soil test can be an important management tool in developing an efficient soil fertility programme, as well as monitoring a field for potential soil and water management problems. Soil samples should be taken before planting. Soil tests provide a scientific basis for regulating available plant nutrients. Recommendations on kinds and amounts of fertilizer to apply and soil management for pepper production are based on the test results. Proper field sampling is extremely important for satisfactory soil reports. A composite sample that accurately represents the field to be tested is most useful. One soil sample for each 2–2.5 ha is sufficient.

When a soil is being tested for nutrient levels and pH, it is a good time to test for nematodes. Nematodes can severely damage pepper roots and reduce yields. Soil samples for a nematode assay should be taken when the soil is warm. Soil samples for nematode assays are taken in much the same way as those for fertilizer requirements. However, these samples must be protected from drying by placing the soil in a plastic bag, by keeping it cool, and by sending it to the assay office immediately. The same agencies that undertake soil testing will usually assay the sample for nematodes.

Soil salinity is also an important consideration when determining how well peppers will grow. Pepper is classified as being moderately sensitive to soil salinity. Average electrical conductivity values of $5\,dS\,m^{-1}$ in the soil profile cause a yield reduction of 50% or more. Lunin *et al.* (1963) demonstrated that the age of the pepper has an effect on the susceptibility to salinity. They found that leaf production dropped sharply when saline conditions were imposed at the early germination stage, while later applications of saline water resulted in only a slight yield reduction. There was also a marked drop in evapotranspiration with an increase in salinity of the water.

PREPARING THE LAND

Crop rotation is an effective way of reducing disease and weed problems in pepper fields. Do not plant pepper in the same field more than once every 3–4 years. Rotation should be to non-solanaceous crops. Crops like wheat, cole crops, maize, lucerne and legumes work well.

Most peppers are grown on soil that is highly prepared with tillage work. 'No-tillage' culture has been examined for peppers by Morrison *et al.* (1973), who found that peppers had lower survival rates in non-tillage plots than either tobacco or tomato. Cultural factors were implicated as being more important in the low survival rate than the physical mechanical transplanting operation. The residual dead sod and previous crop residues are a potential source of plant pests and disease organisms which may attack newly transplanted pepper plants before they become established. However, no-tillage pepper is possible if adequate pest and disease control is provided.

Under a standard tillage method, preparing the soil involves ploughing, deep chiselling, discing, smoothing and listing. An important new technique, laser levelling, which uses a laser to establish the lie of the field, can be advantageous to large field production. Laser levelling the field to a grade of 0.01–0.03% in one or both directions aids in draining the field of extra water, which in turn reduces the risk of root diseases.

Peppers can be grown in a flat field or on raised beds. Raised beds are used in some areas because of furrow irrigation, in others to ensure drainage. Garcia (1908) demonstrated that peppers grown on a high ridge were less likely to have phytophthora root rot as compared with plants grown on the flat ground. For direct-seeded crops, raised beds allow improved control of surface moisture, thereby reducing the chance of infection by soil-borne organisms. Raised beds give protection from root flooding by allowing the root zone to drain after heavy rains. A bed formed by listing soil into ridges is the simplest formation. The ridges are kept moist either with irrigation or rain to establish the field. An option when forming the ridges is to pre-irrigate the field before planting. The seeds are planted into the moist soil and are not irrigated again until they sprout. Another option is to convert two normal width beds into one wide planting bed, called a 'cantaloupe bed' in the United States. The wide bed provides early entrance into a field after rain or irrigation. Because the middle furrow is eliminated, the field does not stay as wet after irrigation.

In cooler climates, the soil can be warmed by orienting the beds in an east–west direction, then sloping the soil bed 30–35° to the south (slant them towards the north in the southern hemisphere). The soil temperature will average 1–3°C warmer than in a traditional flat bed, and increased tillage costs are compensated by better stands and earlier production.

PLANTING

Peppers may be established in the field by direct seeding, by containerized transplants grown in multicellular trays in the greenhouse, or by using bare-root transplants that are field grown. Each method has advantages, and each is suitable for specific production systems. For example, transplanting may result in early production and uniform stands. However, because the pepper field is planted to stand and no extra plants are available, the risk of pests destroying a plant may be a problem. Direct seeding requires less labour and is less costly. However, with new hybrid cultivars costing 10–20 times more than the cost of open-pollinated cultivars, transplanting to a field stand may be the only option.

DIRECT SEEDING

The cost of planting quality seed may only represent 1–2% of the total production cost of the pepper crop. Quality seed represents the basic building block for good pepper production. Whether the seed is used by the grower to direct seed the field or used by a greenhouse grower to start transplants, only the best quality seed should be planted. Seed is the basis for a high-quality pepper crop and poor-quality seed can result in reduction of yield and lower quality of the pepper crop.

When buying pepper seed it is important that certain elements are considered. The most obvious is the cultivar to be planted. Because there are many cultivars within each pod type, selection of the best cultivar for the growing area is important. Past experience with pepper cultivars or advice from a reliable source, such as a university or extension service, are good ways to choose the best cultivar to plant. A respectable seed supplier will also have other important information about the pepper seeds.

Purity is the percentage of pure seed. This purity percentage should be as close to 100% as possible. 'Other crop seed' is given as a percentage of the total weight of the package and is usually 0% in peppers. Inert matter is the non-viable material in the container, such as chaff, soil, broken seeds and sticks. Weed seeds must also be given as a percentage of the total weight. Tolerance for objectionable and noxious weeds is generally zero. Germination percentage is the percentage of pure seed that will produce normal plants when planted under favourable conditions. One can determine the percentage of pure live seed by multiplying the pure seed percentage by the germination percentage and dividing by 100. For example, if a seed lot has 95.5% pure seed and 93% germination, the pure live seed is 88.82%. Thus the value of the seed is determined by the percentage of pure live seed. Some seed companies are now selling their seed by what is termed 'germinable units'. This means that the price of the seed already reflects the pure live seed estimate. The price quoted for seed marketed in germinable units is for 100% pure live seed.

Seeding Rates

Rapid and uniform stand establishment of direct-seeded pepper is important for several reasons: (i) seedlings emerge faster and are less likely to be damaged by soil crusting; (ii) there is less potential for soil-borne diseases and insects to cause catastrophic damage; (iii) rapid root and shoot growth decreases the possibility of desiccation on windy days; and (iv) the crop matures uniformly. Uniform maturity is especially important when mechanical harvesters are used.

Plant population and plant spacing can have a profound influence on plant development, growth and marketable yields. Direct-seeded peppers established by clump planting may require some thinning (see Fig. 7.2) because the competition from more than two or three plants per clump will reduce fruit size.

Sundstrom *et al.* (1984) reported an increase in mechanically harvested tabasco pepper yields as within-row spacing decreased from 81 cm (8200 plants ha^{-1}) to 10 cm (65,000 plants ha^{-1}). Marketable bell pepper yields from populations of 27,000 plants ha^{-1} were significantly lower than those from 40,000 or 60,000 plants ha^{-1}, but the yield increase from narrower within-row spacing occurred only on a two-row bed and not on a three-row bed planting pattern (Batal and Smittle, 1981). In Florida, a spacing of 25 cm within-row spacing with two plants per hill (81,109 plants ha^{-1}) resulted in the optimum plant population for marketable pepper yield (Stoffella and Bryan, 1988). A within row spacing of 30 cm was the best in Texas (Dainello and Heineman, 1986).

Plant architecture can be influenced by altering the plant population. Plant height will increase with higher plant populations. However, Stoffella and Bryan (1988) found that the number of primary or secondary branches generally was not influenced by plant population. The primary and secondary branches are considered to be the location of initial fruit buds and foundation of new fruit bud development in bell peppers. The counts of primary and secondary branches were significantly correlated during anthesis and at final harvest.

Fig. 7.2. Thinning a field of direct-seed peppers.

The increasing fruit number per plant at lower populations was, therefore, not attributed to more primary or secondary branches per plant, but to more fruit per branch. In addition, they found that at higher plant populations the primary branches originated at a higher point on the main stem than at lower populations. Because the fruits are located higher on the plant, lodging at higher plant populations may result, particularly in environments conducive to lodging, for example in areas that are subject to windy and wet weather. Stem diameter generally decreases as plant population increases. Marketable pepper fruit number and weight generally decreased per plant and increased per ha as plant populations increased. Porter and Etzel (1982) reported no significant difference in pepper fruit size between one or two plants per hill spaced 30 cm apart in-row with two-row beds (plant populations of 43,036 vs. 86,072 plants ha^{-1}). However, they did report a fruit size increase in single-row plants with one plant per hill, or populations of 21,518 plants ha^{-1}.

One of the problems associated with high plant populations is when fruit become more yellow during the first harvest, thereby detracting from optimum fruit quality (Stoffella and Bryan, 1988). In addition, Lopez and Silvas (1979) reported a higher incidence of sunscald, lower levels of virus symptoms and greater infection by *Phytophthora capsici* as pepper populations increased from 11,000 to 89,000 plants ha^{-1}.

A kilogram of 100% pure live seed will produce approximately 1 million seedlings. Overseeding is common, and planting at the rate of 2.2–4.4 kg ha^{-1} produces a thick stand. In the southwestern USA, the higher seeding rate is used as a means of controlling curly top disease. The leafhopper that acts as a vector for the disease migrates through the area in early spring, spreading the virus. At the time of thinning, seedlings infected with the virus are rouged, leaving a field stand of healthy plants. This ensures an adequate stand for yield. Growers can also plant to a stand by clumping 3–5 seeds in hills every 15–25 cm along a row.

Enhancement of Crop Establishment

Unfavourable environmental conditions at planting can result in poor plant stands. Complete, uniform and rapid plant establishment under all conditions is a goal of direct seeding. Temperature extremes may cause poor pepper seed germination, resulting in low, non-uniform plant stands. High concentrations of soluble salts in the soil can damage or kill germinating seed and delay emergence. Dry soil conditions can inhibit seedling emergence, while soil crusting, typically caused by heavy rains, may delay, reduce and spread time of seedling emergence.

Seed treatments are available to enhance seed germination and seedling emergence in the field (see Table 7.1). Several procedures should be considered when choosing seed pretreatments. Seed moisturizing brings the moisture

Table 7.1.   Mean rate of emergence (MRE) in the field of tabasco seeds with various treatments. (Source: Sundstrom and Edwards, 1989.)

Treatment	Emergence^z (%)	Stand^y (%)	MRE (days)
Untreated seed, direct seeded	68.5a^x	70.0d	10.4d
GA$_3$-treated, direct seeded	47.2bc	55.0g	9.2c
KNO$_3$-primed, direct seeded	49.4b	68.3e	7.4a
Untreated seed, plug-mix planted	39.7c	66.6f	10.9e
GA$_3$-treated, plug-mix planted	46.4bc	78.3c	8.3b
KNO$_3$-primed, plug-mix planted	53.3b	81.7b	7.6a
Pregerminated seed, plug-mix planted	66.1a	90.0a	7.5a

^{z}Emerged seedlings per hill per seeding treatment.
^{y}Number of hills containing at least one seedling.
^{x}Means within a column followed by different letters are significantly different at $P \le 0.05$.

content of the seed to 10–25%. The seed is then handled as dry seed and planted by conventional methods. This method is used to avoid damage that sometimes occurs from rapid water absorption.

Pre-germination is a technique where seed is brought up to a moisture content between 30 and 50%, suspended in a gel, then extruded behind the furrow opener of a planter. These seeds must be handled with extreme care and should not be allowed to dry out. Research shows that pre-germinated pepper seed, planted by fluid drilling techniques, emerges 3–4 days sooner than dry seed, but fluid drilling requires special planting equipment. There are some problems associated with this method. Root radicles are already emerged from the seed coat at the time of planting. If the radicles are longer than 4 mm, they can be damaged during planting. This method is not recommended in areas with cool soil temperatures, as food reserves in the seed are depleted faster in cool soils and may prevent emergence completely.

Fluid drilling of germinated seed results in earlier, more uniform emergence, better plant stands, and superior seedling performance than sowing seeds dry, especially at low temperatures. However, stands from fluid-drilled seeds have been reduced when sown under high temperatures. Another direct seeding method is plug-mix seeding, which incorporates dry or germinated seeds into a moistened medium containing peat, vermiculite, fertilizer and lime (Schultheis *et al.*, 1988a). Pepper crops are difficult to establish by plug-mix seeding during periods of high temperature and dry planting weather. The plug-mix dries quickly and desiccates emerging seedlings unless watered immediately. Heavy rains also tend to wash seeds from the plug-mix media. Schultheis *et al.* (1988b) combined fluid-drilling and plug-mix seeding systems and called it 'gel mix seeding'. The combined system had the benefits of each system: the use of germinated seeds, anticrustant properties, reduced pathogen microenvironment, added nutrients, high water holding capacity and maintenance of a high-moisture environment. They concluded that a gel mix of 1.25-l gel solution at 1% Liqua-gel (g l^{-1}) for each litre of plug-mix led to early, uniform and complete plant stands under varying environments.

A less difficult method is seed priming. Seeds are soaked for 3–7 days in a salt solution or a solution of polyethylene glycol (PEG). The seed begins the germination process while it is soaking in the priming solution. Priming is stopped just prior to radicle emergence from the seed coat. The seed is then dried down to its original moisture content and planted as a dry seed. Primed seed germinates faster and more uniformly than non-primed seed, especially in cool soils. Seeds should be stored under optimum conditions, or the effectiveness of seed priming can be reversed. Some seed companies will sell primed seed if orders are placed far enough in advance for them to treat the seed. One disadvantage of seed priming is that variation of soil temperature makes results unpredictable from one year to the next. Seed priming is becoming more popular in the transplant industry, where uniform germination in the greenhouse is critical.

A technique used in furrow irrigated pepper production is to 'cap' the seeds. This technique places the seed in moist soil and then places a 7–10 cm high soil cap on top of the seed to reduce water evaporation. When seeds begin to germinate (the crook stage), the cap is removed with a dragging harrow. The soil-removal equipment is carefully adjusted so a loose, 0.5 cm layer of soil covers the seedlings after dragging. This avoids seedling damage and encourages rapid seedling emergence.

Seed can be planted either in the centre or on the edge of a shaped bed, depending on the soil's salt content. Pepper seedlings are susceptible to salt until they are 5–8 cm tall and growing rapidly. If soil has high residual salt, or irrigation water is salty, plant to one side of the bed to reduce salt concentration near seedlings. When planting on the side, move soil from the adjacent bed during cultivation. Continue to move soil until the plant row is eventually centred and on a high ridge. When the 'cantaloupe bed' method is used, the seeds are planted on each edge of the wide bed. Water in the furrow wets both edges of the rows, while salt accumulates in the centre of the bed, away from both seeded rows.

With overseeding, the pepper field must be thinned. This should be done when pepper plants are 5–10 cm tall, with 3–4 true leaves. Thinning should be done when the chance of natural damage that may kill plants, such as curly top virus, damping-off and salt injury, has passed. Research has shown that good yields can be obtained from single plants, or clumps of no more than three plants, uniformly spaced 25–30 cm apart in the row. Depending on row spacings, a plant stand ranges from 32,000 to 100,000 plants ha^{-1}.

In the southwestern USA, the most common row width is 0.8–1.0 m, a carry-over from cotton growing days. Growers often select row widths to conform to requirements of other farm crops. Narrow row spacing, 80 cm, can result in higher yields, particularly for a second green harvest or for a late red harvest, but may also increase the danger from diseases, e.g. *Phytophthora capsici*.

Soil crusting by rain can cause a problem when peppers are direct seeded and have not emerged. Frequent irrigation prior to emergence is one remedy, but this produces unnecessary increases in water use and production costs. Anticrustants, if effective, could alleviate these additional costs. Anticrustants reduce hypocotyl stress of germinants, but do not affect stand establishment, growth or yield of peppers (McGrady and Cotter, 1984).

TRANSPLANTING

Pepper seeds can be sown directly in the field, but many producers transplant young seedlings. The term transplanting means to move a plant from one soil or culture medium to another. Transplanting peppers is a suitable option because of slow seed germination. Pepper seed can take up to 3 weeks to emerge. Pepper transplants are ready in 6–8 weeks from seed. By using transplants, fields can be maximized for production. Transplanting has some advantages. It helps guarantee a well-distributed stand of plants, reduces seed costs and thinning costs, and requires less cultivation and irrigation. A crop can usually be established with one irrigation, compared with as many as three irrigations for directly seeded stands. Such economies help offset transplant and field setting costs.

Pepper plants established from transplants are more uniform, can tolerate or escape early abiotic and biotic stresses, and may achieve earlier maturity than direct-seeded plants. The choice of planting system depends on the economics of plant establishment, plant performance and establishment and, most importantly, the value of the subsequent yield. Transplants can also promote earliness or allow for later than normal plantings. There is also the possibility of increasing the number of harvests, thus increasing yield per unit area.

Given these advantages, several factors should still be considered when making the decision to direct seed or transplant peppers. Soil and air temperatures, along with seed price are usually the most important factors. Other factors such as the use of mulches and drip irrigation may make using transplants more enticing. Anticipated benefits of earliness and higher yields are not consistent, so growers should consider other factors (seed amount, thinning costs, water amount and late planting opportunities) when deciding whether to transplant. The price of seed versus the price of transplants, the labour availability and cost, and the availability and cost of automated equipment are three factors to consider in overall cost comparisons. The amount of water required to establish seeds versus transplants should also be considered.

In many production areas transplants are started in greenhouses or hotbeds, or in mild climates, in outdoor seedbeds 6–8 weeks prior to field planting. Peppers are transplanted when they are 6–8-week-old plants and

15–20 cm tall. Also, stem diameter is very important to survival rate in transplants. The thicker the stem the higher the survival rate. Applying a high-phosphorus starter solution to the soil during transplanting aids establishment. Prior to field planting, pepper transplants should be hardened but not excessively. Plant growth can be delayed substantially by over-hardening.

A distinctive characteristic of direct-seeded plants versus transplanted pepper plants is the root morphology. Direct-seeded plants have a strong taproot whereas transplants, because of the early containerization, lose the taproot and form extensive lateral roots. Transplants have lower total root growth than direct-seeded pepper plants, but produce earlier and higher fruit yields (Leskovar and Cantliffe, 1993). Transplants are shorter than direct-seeded plants and have more branches. This can be detrimental when long fruits touch the soil, increasing the possibility of pod rot.

PLANT SPACING

Whether the field is transplanted or direct seeded, the optimum plant population is not only dependent upon the in-row spacing and between-row spacing of the plants, but also by the plant type grown. For example, spacing of bell pepper plants will differ from the spacing of jalapeño, New Mexican, or pepperoncini pod types. Many studies have been published on the optimum plant population for bell peppers. Miller *et al.* (1979) reported that bell pepper grown at a plant population of 48,000 ha^{-1} resulted in a relatively low leaf area index (LAI = leaf area ÷ ground area covered) at 98 days after transplanting. They suggested that bell pepper efficiency increases at higher plant populations. Motsenbocker (1996) found similar results with pepperoncini peppers. In general, plants grown at the narrowest spacings produced the smallest plants, leaf and stem biomass, but resulted in a more upright plant, and produced the highest fruit yields and counts per hectare and the lowest fruit yield per plant. The higher plant population at the narrower in-row spacings compensates for lower yield per plant considering that fruit width and size are unaffected and result in increased yield per area. Cotter (1980) studied plant spacing for the New Mexican pod type and found that plants spaced at 35–46 cm within the row and 1 m between rows was the optimum.

MULCHES

Plastic (polyethylene) mulch (see Fig. 7.3) has been used on peppers since the early 1960s. The advantages are increased early yields, aids in moisture retention, inhibition of weeds, reduced fertilizer leaching, decreased soil compaction, fruit protection from soil deposits and soil microorganisms, and facilitation of fumigation. The three main 'colours' of mulch are clear, black and coated. The

Fig. 7.3.   Bell peppers on a wide mulched bed in Florida, USA.

coated mulch can be painted or tinted to almost any colour (Maynard and Hochmuth, 1997).

Plastic mulches are often used in conjunction with drip irrigation when establishing transplants. Plastic mulches have been shown to increase soil temperatures, hasten maturity, increase yields, increase produce quality, and help control insects and diseases. Plastic mulch functions as a barrier to the packing action of heavy rainfall and reduces the tendency of workers to walk on the drill area. Mulches are an excellent soil cover and can increase the effectiveness of soil fumigants by reducing the escape of gaseous fumigants and facilitating more uniform distribution. Increased yields of bell peppers have been reported with plastic mulch often in combination with soil fumigation and trickle irrigation. The initial cost of using plastic mulch is high, as special equipment is needed to lay the plastic and drip lines. This initial cost is offset by higher yields, reduced weed control and increased water efficiency. Disposal of the plastic at the end of the season can be an environmental concern.

One of the main objectives in using plastic mulch is to increase soil temperature in the root zone. The favourable temperature promotes better root growth which in turn promotes better foliage growth and fruit set. Black mulch is the most common colour used (see Fig. 7.4). Higher soil temperatures are possible under clear plastic, but weed growth under clear plastic is often a serious problem. Clear plastic transmits a high percentage of the light energy which in turn is converted to heat in the top layer of soil. Black plastic is opaque and transmits almost no light rays. Therefore, the light rays are changed to heat energy in the plastic film itself. The heat from the plastic is absorbed by the soil and is released into the air. The most efficient movement of heat into the soil is where the plastic is in contact with the soil.

Fig. 7.4. Peppers grown on a raised bed with black mulch.

Plastic mulch can affect the microclimate to change the growth and development of pepper plants. The mulch surface colour affects the growth and development of bell pepper plants. In areas where late summer or autumn plantings are possible and soil warming is not beneficial, a white-surfaced mulch is often used. Pepper plants growing over red mulch are taller and heavier as compared with plants grown over black and yellow mulch (Decoteau *et al.*, 1990). Mulch colour did not affect leaf area per plant. Darker colours of mulch (red and black) reflected less total light and more far-red and red light and also warmed the soil more than yellow and white (Decoteau *et al.*, 1990).

White mulches (see Fig. 7.5) can modify the radiant energy levels entering a foliage canopy by increasing the soil surface reflectance. Gerard and Chambers (1967) reported that reflective coatings increased the yield of bell peppers over irrigated bare ground plots. Dufault and Wiggins (1981) reported that plants grown over white mulch were shorter, fruited earlier, and produced higher overall yields than unmulched plants. Reflective mulches increased earliness and yield. They found that solar reflectors constructed in the field had little effect on pepper growth and yield.

Inhibition of weed growth is attainable when opaque plastic mulch film is used. Clear mulch can accentuate the weed problem in cool areas. Herbicides

Fig. 7.5. Pepper grown on a raised bed with white mulch.

registered for pepper cultivation will control weeds in the unmulched middle furrow. Very little fertilizer leaching from rainfall occurs in plastic mulched beds and pepper roots will be confined almost exclusively to the area under the mulch. By applying all the fertilizer in the soil under the mulch, a reduction in total fertilizer per season may be realized. Higher quality and a lower percentage of cull grade fruit are usually achieved in well-managed plastic mulch culture.

Mulches have also been shown to reduce aphid infestation and the concurrent virus damage associated with aphid transmission of viruses. On peppers, Porter and Etzel (1982) noted yield increases with mulches, particularly reflective mulch, and the potential of mulches in reducing the incidence of virus disease on peppers was demonstrated by Black and Rolston (1972). Bell peppers grown on a silver reflective plastic mulch produced greater yields than plants grown on black plastic or bare ground. This increase was found even when aphid-borne viruses were not an apparent problem. The increased yields were attributed to the increased light (photosynthetically active radiation) reflection of the aluminium painted polyethylene. The number of aphids trapped over the aluminium painted mulch plots was less than 10% of those trapped over rows with black polyethylene or no mulch during the first 3 weeks after planting. During the next 5 weeks, the percentage gradually increased to about 50%. At the first harvest, 10% of the plants grown on aluminium mulch were showing mosaic symptoms as compared with 85% of the plants on black plastic and 96% on no mulch plots. Plants in aluminium mulch plots yielded 58% more than those on black polyethylene, and 85% more than those on no mulch. Black and Rolston (1972) also looked at tabasco plants and found that the number of plants killed by tobacco etch virus was lower on the aluminium mulch. On aluminium mulch 42% of the plants were killed as compared with 96% for black polyethylene mulch and 98% without mulch.

When plastic mulch is used with drip irrigation the total volume of water needed for a pepper crop may be reduced by as much as 50% compared with overhead sprinkler irrigation. However, the use and adoption of reflective mulches is limited by some of its disadvantages. Primarily, reflective mulches are relatively expensive compared with black plastic mulch, and secondly, they result in reduced soil temperatures, which can reduce their potential for use in early spring pepper production.

When plastic mulch is used in pepper production, fields must be selected carefully. Avoid using plastic in weed-infested fields, unless fumigation is planned. Apply a 'full-season' amount of phosphorus, potassium, and minor nutrients and 30–50% of the nitrogen before placing the plastic on the beds. Once the mulch is laid, it is very difficult to side-dress low solubility nutrients effectively during the growing season. Most nitrogen fertilizers are readily soluble in water and can be side-dressed through a drip system.

The mulch should be placed when the soil of the planting beds is well pulverized and moist to facilitate water and heat energy management and efficiency. Lay the plastic so that it fits snugly over a smooth, level surface to get the maximum plastic-to-soil contact. The plastic should have a 'j' shaped tuck on both sides after being laid. This will greatly reduce the likelihood that wind will whip the plastic loose from the bed. Bury the irrigation tape about 2.5 cm deep with emitter holes facing up. The best results are obtained when the tape is put down when the bed is being made. When drip tape is placed on top of a bed instead of being buried, it tends to twist and slide under the plastic during a few days of heating and cooling. This makes uniform watering of beds almost impossible.

Before planting the pepper transplants, the irrigation system should be hooked up and tested. The warmed soil immediately surrounding a planting hoe can dry out quickly and short watering periods may be required even before the pepper plants are planted. Soil moisture under mulches should be checked daily. Determining when and how much water to apply will be among the most difficult and critical production management decisions to make. If a bed becomes completely dry, it can be very difficult to moisten again. When transplanting, the hole in the plastic should be as small as possible. Weeds will grow if the hole is too large. A common method to make the holes efficiently is to use a hand-held propane torch to burn a hole.

When rice straw (see Fig. 7.6) and plastic mulches were tested in the tropical lowlands of Indonesia, they affected the soil temperature, light reflection and soil nutrient concentrations after the last harvest (Vos and Sumarni, 1997). Rice straw mulch reduced soil temperature, induced faster plant growth, advanced mid-fruiting time and resulted in higher potassium content in leaves, but had no effect on crop production. Plastic mulches increased soil temperature, induced faster plant growth and earlier fruiting,

Fig. 7.6.   Peppers being grown with straw mulch in Korea.

reduced phosphorus concentration in leaves and fruits, and increased nitrogen concentration in leaves and fruits. Yield and mean fruit weight of healthy fruits were increased and earliness of harvesting was enhanced. Improved crop performance and production with increase of fertilizer efficiency, control of evaporation, leaching and soil erosion makes plastic mulch the best for peppers.

Pepper needs relatively high soil temperatures (25–31°C) for optimal germination and emergence. Direct sowing in the field in spring can mean planting in cool soils. Direct seeding under clear polyethylene mulch can increase soil temperature to improve seed germination. In addition, clear polyethylene can reduce soil crusting. Cavero *et al.* (1996) found that plant stand was improved from 0% for unmulched to 60% for the clear polyethylene mulched direct-seeded rows. Once the seedlings have emerged, the clear polyethylene is removed from the field.

Peppers have maximum shoot dry weight and leaf area when the root-zone temperature is 24°C and 30°C respectively (Gosselin and Trudel, 1986). Leaf area ratio was not affected by root-zone temperature, but fruit weight was maximum at 30°C. Thus mulching would aid pepper emergence in areas where low soil temperatures are present early in the season. A drawback to clear polyethylene plastic is a soil temperature too high for seedling growth. The plastic needs to be removed before temperatures become too high.

Albregts and Howard (1973) examined the response of peppers to the use of paper mulch coated with a thin layer of polyethylene. Unlike polyethylene, the paper mulches are biodegradable and will disintegrate after incorporation into the soil. One of the principal benefits of mulch on sandy soils is the reduction in fertilizer leaching. Strip mulching, the use of a narrow strip of mulch placed over the fertilizer, has also shown promise in reducing leaching. They concluded that early growth and seasonal marketable yield was increased using the mulch as compared with no mulch. Both the whole bed and a strip mulch increased yield of peppers. Pepper fruit size and number were increased by mulch treatment. As amount of bed covered by mulch increased so did the soil nitrogen and potassium.

Gerard and Chambers (1967) indicated that reflective paint sprayed on mulches could help with pepper seed germination in those areas where soil temperatures are too high for pepper seed germination (greater than 43°C). In warm areas, autumn-planted peppers may not germinate because of high soil temperatures. Thus if a coating could be sprayed over the planted seeds to lower the soil temperature, this could increase plant stand and directly increase yield. They sprayed a 15 cm strip of a white to cream coloured petroleum resin emulsion over the planted pepper seeds. The coating was effective in suppressing soil temperatures to below 43°C at the planting depth and thus provided a favourable environment for germination of pepper seeds. The material also suppressed soil moisture evaporation ensuring that the seeds received enough moisture to germinate.

Soil Solarization

Clear plastic mulch has been used for soil solarization of pepper fields. Soil solarization is the use of the sun to heat soil during a fallow period. Solarization generates temperatures sufficient to control many annual weeds and soil-borne pathogenic fungi. It is useful in areas that have hot dry periods. Solarization increased marketable yield of autumn-grown bell peppers by 20% in Texas (Hartz *et al.*, 1985). The clear plastic can be sprayed with paint and left in place as a mulch. When this was done by Hartz *et al.* (1985) they realized a 53% increase in yield. The soil solarization and painted mulch did not influence the earliness of the pepper crop.

There are disadvantages in using mulches. Removing and disposing of the used plastic mulch is the biggest problem associated with mulches. Non-degradable mulch breaks down slowly in continuous exposure to summer sunlight. Plastic not exposed to sunlight, i.e. those parts covered with soil to hold the plastic in place on the bed, do not break down. These remnant pieces can become a serious problem in soil preparation, seeding and cultivation. Degradable mulches can be either biodegradable or photodegradable. Photodegradable means that the plastic will degrade after a specific number of hours of sunlight (ultraviolet rays). Biodegradable mulch does not need the sunlight for degradation. Biodegradable mulches formulated to break down after a period of time are preferable.

Another disadvantage of plastic mulch is that specialized equipment is needed when using it. Several attachments are needed to lay the mulch, shape the beds, place the fertilizer and install the drip irrigation tape. In addition, fumigant injector attachments along with other transplanting aids can become expensive for the farming operation.

IRRIGATION

In areas with regular and ample rain, irrigation is not needed. Irrigation is essential in arid and semi-arid regions to provide adequate moisture for production of peppers. Today, many hectares of peppers are grown with irrigation in the semi-arid regions of the world. Although there is evidence that pepper plants are drought resistant, fruit set tends to be depressed by any extreme environmental condition. Irrigation is essential for pepper production, as peppers can require up to 60–75 ha-cm of water during the growing season. Peppers are known to be sensitive to moisture stress at flowering and fruit setting. If plant growth is slowed by moisture stress during blooming, blossoms and immature pods are likely to drop. Blossom-end rot can result when plants are stressed when young fruit are developing rapidly. Water-stressed plants generally produce more pungent pods.

Pepper is a shallow-rooted crop so the amount and frequency of irrigations depends on soil type, bed type, plant size, humidity, wind, sunlight and prevailing temperatures. Up to 70% of the water absorbed by full-canopy peppers is removed from the top 30 cm of soil. Reduced yields and decreased fruit size will result from moisture stress. In Ethiopia, it is reported that the full potential of the pepper crop has not been fully exploited due primarily to the lack of irrigation (Haile and Zewdie, 1989). Limiting the water applied to pepper during the period of rapid vegetative growth reduces the final yield (Beese *et al.*, 1982).

Optimum irrigation time can be determined by testing the soil moisture in the root zone by touch, with moisture sensors or by computer predictions. When estimating when to water by watching the pepper crop, the leaves are the best indicator. During hot, drying conditions, expect swiftly growing plants to wilt late in the afternoon, even on the first day after irrigation. Wilting signs begin to appear earlier in the day as soil dries. When plants wilt in the early afternoon, irrigate. Water may have to be applied on a 5–7-day schedule in summer to prevent blossom-end rot. Decreasing irrigation frequency at the end of the season will promote ripening and improve red fruit colour.

Excessive irrigation can be as harmful to the crop as too little water. Phytophthora root rot disease can develop from water standing in the field for more than 12 h, so a means of draining the field is helpful. Frequent light irrigations are better than infrequent heavy irrigations with peppers because of their shallow roots. Drip (trickle irrigation) is one method of application to optimize the water for pepper production and to conserve water in arid regions. Drip irrigation with intensive cultural practices such as the use of mulches generally results in additional yield increases. Drip irrigation allows for frequent application of low levels of soluble nutrients to the root zone. The additional control obtained over the root environment is a major advantage over other irrigation systems. However, when drip irrigation was compared with sprinkler irrigation in a humid climate, i.e. New York, both worked equally well as long as the water application was based on soil matric potential (VanDerwerken and Wilcox-Lee, 1988). It may be that the increased yields reported due to trickle irrigation in humid areas are due to the cultural practices often accompanying the use of trickle, such as mulch and timely fertilizer and pesticide applications through the drip system.

Fertigation is the application of fertilizer through the drip-irrigation water. Keng *et al.* (1979) reported that fertigation helped increase pepper yields in the Oxisols of the wet-dry tropics and was superior to broadcasting nutrients. A high percentage of soils in tropical countries are Oxisols. They are character-ized as having low cation exchange capacities and contain high levels of sesquioxides and kaolinite and related clay minerals. Because Oxisols have low nutrient and available water holding capacities, drip irrigation helps maintain a favourable moisture and nutrient level of the root zone.

CULTIVATION

As soon as young pepper plants become established in the field (2.0–5.0 cm high) a shallow cultivation will control weeds. However, deep cultivation, as done on cotton crops is not recommended for peppers. The deep cultivation can wound the pepper roots and can result in a higher incidence of root diseases.

FERTILIZERS

Peppers require adequate amounts of most major and minor nutrients. The nutrients normally used most on peppers are nitrogen and phosphorus. Pepper appears to be less responsive to fertilizer, especially phosphorus, than onions, lettuce and cole crops (Cotter, 1986). The first nitrogen application and all the phosphorus can be broadcasted before discing or listing the field. Alternatively, phosphorus can be banded at 15 kg ha^{-1} 8–10 cm below the seed. This is a more efficient method of applying the phosphorus.

Pepper benefits from some nitrogen, but too much nitrogen can over-stimulate growth resulting in large plants with few early fruits. During periods of high rainfall and humidity, excess nitrogen delays maturity, resulting in succulent late-maturing fruits and an increased risk of serious plant or pod rots. Nitrogen requirements for pepper production have been studied extensively (Maynard *et al.*, 1962; Stroehlein and Oebker, 1979; Batal and Smittle, 1981; Locascio *et al.*, 1981; Hartz *et al.*, 1993). Substantial differences in nitrogen response have been reported with maximum productivity achieved with rates from 70 kg ha^{-1} to greater than 200 kg ha^{-1}. This large disparity is due to regional and seasonal differences in environment and cultural practices which affect plant vigour and nitrogen availability and uptake efficiency. The interactions of nitrogen rate, application timing and factors controlling nitrogen leaching losses are particularly difficult to reconcile. Payero *et al.* (1990) found that 240 kg ha^{-1} of nitrogen gave the highest yield, while Panpruik *et al.* (1982) found no differences in yield from 0 to 224 kg ha^{-1} of nitrogen. Some commerical pepper producers use greater than 300 kg ha^{-1} to ensure vigorous plants and maximum yields.

Preplant nitrogen also generates vigorous seedling growth, which ensures a well-branched plant by the first fruit set. Stroehlein and Oebker (1979) concluded that moderate rates of nitrogen, 100–150 kg ha^{-1}, produced a more desirable plant and highest yields. Cotter (1986) recommended that 22–34 kg ha^{-1} be broadcasted before discing. Liquid forms of nitrogen may be banded at the rate of 2–6 kg ha^{-1}, 10 cm below the seed. Preplant nitrogen is not needed if a soil test shows the soil has 20 ppm nitrate or more. Cochran (1932) pointed out that moisture and nitrogen nutrition are among the essential factors influencing reproduction development in peppers. Miller (1961) concluded that improved fruit quality with high nitrogen levels was related to a steady increase in nitrogen content of plant and fruit tissue.

Nitrogen content of pepper leaf tissue has been used to monitor the nitrogen status of pepper production. Petiole nitrogen levels are not as varied as those in soil samples. The University of California recommends NO$_3$-N for bell pepper to be 10,000, 5000 and 3000 µg g^{-1} of tissue at early growth, early fruit set and late fruit bulking period, respectively (Lorenz and Tyler, 1983). Most published values agree with this level. Hartz *et al.* (1993) state that a concentration of greater than 5000 µg g^{-1} through the early fruit setting stage will maximize fruit yield. Batal and Smittle (1981) reported that the highest marketable yield resulted when sufficient nitrogen was added to maintain soil NO$_3$-N levels between 20 and 30 ppm for spring- and autumn-planted peppers, respectively. Their study in Georgia, USA found that yield increases were influenced by frequent irrigation only when additional nitrogen was applied to maintain a higher soil NO$_3$-N level. The number of additional nitrogen applications were about doubled to raise the soil NO$_3$-N level from 10 to 20 ppm or from 15 to 30 ppm. Soil moisture and stage of plant growth affected the nitrogen content of the sampled leaf tissue. They suggested that both soil and leaf tissue analysis should be used to determine nitrogen applications. Hartz *et al.* (1993) reported that analysis of fresh petiole sap by nitrate-selective electrode was highly correlated with dry petiole NO$_3$-N level and appears to be a viable, on-farm technique to supplement conventional laboratory tissue testing.

Slow-release fertilizers extend nitrogen availability and reduce nitrogen losses. Slow-release methylene urea, sulphur-coated urea and ammonium sulphate fertilizers were tested on peppers by Wiedenfeld (1986). Methylene urea and sulphur-coated urea may improve pepper yields by improving nitrogen availability, thus reducing the rate and number of nitrogen fertilizer applications required. He concluded that the extra cost of slow-release fertilizers was not justified. A single nitrogen application of a soluble nitrogen source was the best method of supplying the nitrogen requirement of peppers in this study. If numerous split applications are used to ensure that nitrogen availability does not limit subsequent crop growth, a single early application of a slow-release fertilizer may accomplish this objective efficiently. Nitrogen availability to the young seedling appears to be especially important, because a preplant application of a soluble nitrogen fertilizer performed just as well as slow-release materials.

Thomas and Oerther (1972) estimated nitrogen content and therefore the requirement of pepper plants by using spectrophotometric measurements. Light absorptance by leaves in the visible region of the spectrum depends primarily on the concentration of the chlorophylls and carotenoids. Because a deficiency of any one of several nutrient elements decreases pigment formation and subsequent leaf colour, limiting these elements would increase reflectivity because of decreased radiation absorption. They were able to quickly estimate the plant's nitrogen status by using spectrophotometrically measured diffuse reflectance from the adaxial leaf surfaces of pepper. Because the spectrophotometric measurements detect changes in leaf colour before human visual detection, the nitrogen status of plants can be measured in the field instead of the laboratory.

MYCORRHIZAL FUNGI

Vesicular-arbuscular mycorrhizal (VAM) fungi (*Glomus macrocarpum, Glomus deserticola, Glomus mosseae, Glomus intraradices,* etc.) increase nutrient uptake, especially of phosphorus, and growth of pepper plants in nutrient-poor soils. There can also be an effect in fumigated soils when the VAM fungi are killed along with the pathogenic fungi. The lack of a VAM-associated root system is usually expressed by poor growth resulting from phosphorus deficiency in the plant. If this deficiency is not great, the lack of optimum growth may not be recognized at all. VAM pepper plants when grown in nutrient-poor soils had an increase from 44 to 188% over the non-VAM plants (Haas *et al.*, 1986). VAM colonization begins as early as 3 days after germination. After 21 days, 60% of the seedling population had VAM associations (Afek *et al.*, 1990).

In field situations, roots may be infected with VAM by secondary infection along the roots. In normal root systems, the young, potentially colonizable roots are produced as branches from other roots. In commercial operations, inoculating peppers with VAM in the nursery or greenhouse would be an acceptable method, instead of inoculating the field.

NUTRIENT DEFICIENCIES

Several other nutrients have been reported to cause a loss of pepper yield when there is an inadequate amount during the growing period. Table 7.2 shows the target levels of various elements required for healthy growth in sweet peppers. Miller (1961) found that peppers deficient in phosphorus produced weak plants. The leaves were narrow, glossy and developed a greyish green colour. The red or purple coloration of stems and leaves often associated with phosphorus deficiency did not develop on peppers. Fruits produced on pepper plants deficient in phosphorus were shorter, narrower and resulted in a pointed tip on bell peppers. Plants with a phosphorus content of 0.09% or lower in vegetative tissues exhibited deficiency symptoms.

Ozaki and Hamilton (1954) described a bronzing condition of pepper leaves, followed by necrosis and leaf drop associated with low levels of potassium. Miller (1961) found the same symptoms and, later, small necrotic lesions developed along the veins, followed by defoliation. Symptoms of potassium deficiency were associated with 1.17% or lower levels of potassium in the vegetative growth. Low calcium levels produced stunted plants and severe blossom-end rot.

Magnesium deficiency in pepper is characterized by pale-green leaf colour followed by interveinal yellowing, leaf drop, small plants and undersized fruit. Magnesium deficiency generally occurs in pepper plants grown on acidic, sandy soils in areas of high rainfall. Magnesium deficiency may be prevented by soil applications of magnesium sulphate (Epsom salts) or foliar application of magnesium salts. Spray applications of magnesium and other minor elements (e.g. iron) are more effective, rapid, but of a shorter duration as compared with soil application. Magnesium deficiency causes interveinal chlorosis while the

Table 7.2. Target levels for tissue analysis for sweet peppers. (Source: Portree, 1996, *Greenhouse Vegetable Production Guide.*)

Element	Normal range	Deficiency
Nitrogen	3.5–5.5%	< 2.0%
Phosphorus	0.35–0.8%	< 0.2%
Potassium	3.0–6.0%	< 2.0%
Calcium	1.5–3.5%	< 1.0%
Magnesium	0.35–0.80%	< 0.30%
Boron	30–90 ppm	< 20 ppm
Iron	80–200 ppm	< 60 ppm
Manganese	100–300 ppm	< 20 ppm
Zinc	40–100 ppm	< 25 ppm
Sulphur	0.37%	n/a
Molybdenum	6–20 ppm	< 4 ppm

veins and a narrow adjacent portion of the leaf surface remain green. Necrotic lesions develop later in the chlorotic areas, particularly in the upper portion of the plant.

FLOWER DROP

The tendency of the bell pepper to drop many of its flowers and small fruits, resulting in a low early season yield and a corresponding loss of market advantage is a recognized problem in the United States. Wien *et al.* (1989) reported that cultivars differ in their susceptibility to stress-induced flower abscission (drop). They found that susceptible cultivars reduce assimilate partitioning to flower buds and maintain high assimilate consumption of expanded leaves. However, the preferential partitioning of assimilate to young leaves did not appear to be involved in flower drop between susceptible and resistant cultivars.

GROWTH REGULATORS

Many growth regulators have been reported to affect peppers. The most common or most frequently studied include gibberellic acid, ethephon and indoleacetic acid.

Abnormalities in pepper flowers caused by gibberellic acid were reported by Sawhney (1981). Young pepper plants treated with gibberellic acid (GA$_3$) before the initiation of floral organs produced abnormalities in subsequently formed flowers. The development of petals and stamens was affected by GA$_3$, but sepals and the gynoecium were insensitive to the treatment. The effect on petals was limited to the unrolling of the petals, but stamen development was affected more dramatically. GA$_3$ caused abnormalities in pollen development and induced the carpelization of stamens. The expression of feminization of stamens ranged from the production of a few external ovules to a complete transformation of a stamen to a carpel with ovary, style and stigma. In some instances, the growth of stamens was also inhibited. GA$_3$ also induced supernumerary organs in flowers, all of which were 'carpel-like'.

A triazole growth regulator (uniconazole) was tested on potted ornamental pepper for its response on plant height and fruiting (Starman, 1993). Foliar spray concentrations from 5.0 to 15.0 mg l^{-1} gave adequate height control. However, 15.0 mg l^{-1} reduced height excessively when applied at 8 weeks but not at 10 weeks after sowing. An increase in the percentage of red fruit was seen with an increase in the concentration of uniconazole when applied at 10 weeks, but not at 8 weeks after sowing.

Ethephon has been tested on ornamentals as a growth regulator to hasten ripening and to control plant height. Ethephon applied as a foliar spray at

300 ppm increased the number of lateral branches, but delayed flowering and reduced fruit production (Khademi and Khosh-Khui, 1977). Concentrations of 150 and 300 µl l⁻¹ were effective in accelerating fruit ripening of ornamental peppers (Armitage, 1989). It was also observed that concentrations as low as 75 µl l⁻¹ were effective, but a concentration of 600 µl l⁻¹ resulted in foliar and fruit damage. Fruit that were less than 3 cm long were less sensitive to ethephon than more mature fruit. When the pH of the solution of ethephon was raised from 3.3 to 6.3 the treatment effect was increased.

Indoleacetic acid and benzyladenine have been tested for their effect on peppers. When they were tested on ornamental peppers for their effect on lateral branching, concentrations as high as 150 ppm for indoleacetic acid and 1200 ppm for benzyladenine did not increase the lateral branching of potted ornamental peppers (Khademi and Khosh-Khui, 1977).

WEED CONTROL

Competition between weeds and peppers for nutrients, light and water is a serious problem in pepper production (Lee and Schroeder, 1995). Severe weed pressure may reduce yields, impede harvesting operations and clog machinery. A successful weed control programme is essential for producing a healthy crop of peppers. Because pepper cropping systems differ from one area to another, weed management strategies will also differ.

Weeds have an impact on pepper production in several ways. Direct-seeded peppers emerge slowly from the soil and continue to grow slowly, making them more susceptible to competition from weeds for sunlight, nutrients, water and space. Weeds emerging after thinning can reduce yields even if fields are kept clean prior to thinning. Weeds in the field at the end of the growing season interfere with the harvesting process, making harvest more expensive and difficult; additionally weeds will have produced seed allowing for them to return next year. Weeds also serve as alternative hosts for other pests, including nematodes, insects and viruses.

There are hundreds of plants that can be weeds in pepper fields. A listing of weeds found in pepper fields worldwide is beyond the scope of this book. Nevertheless, the first step to weed management in pepper production is to identify the weed correctly. It is also important to understand the life cycle and reproductive capacity of the weed. Life cycle information includes germination, vegetative growth, flowering, seed set and death. It is best to view weeds as either grasses or broadleaves, then fit them into the categories of annuals, biennials and perennials.

Weeds that germinate and complete their life cycle within 1 year are termed annuals. They spread and reproduce only through seed production. Depending on climate, there can be summer annuals, which germinate throughout the spring and summer and complete their life cycle in the autumn of the same year, and winter annuals, which germinate in the autumn of the year, overwinter and complete their life cycle in the spring. Biennials are plants that require 2 years to complete their life cycle. They spread and reproduce by seed production only. Perennial weeds live for 2 or more years. They reproduce by seed and vegetative reproductive structures, such as root buds, rhizomes, crowns, tubers, stolons and bulbs.

As mentioned earlier, mulching can reduce weeds in the field. Where mulching is not used, weeds in pepper fields are controlled for the most part by tractor cultivation and hand hoeing. The use of cultivation as a weed management tool is an ancient practice and is still quite effective in managing annual weeds if done when the weeds are small. With perennial weeds such as Johnson grass (*Sorghum halepense* L.), using cultivation for management often aids in spreading this plant by breaking up and moving around underground vegetative reproductive structures. Cultivation can be used to provide some weed management within the row by moving soil around the base of the plant when the pepper plants are around 12 cm tall. This layer of soil will prevent some weeds from emerging by changing the soil environment around the pepper plant from one favourable for weed germination to one that is unfavourable. To eliminate weeds between pepper plants within the row usually requires hand hoeing. The expense of hand weeding can be reduced by close cultivation before thinning, as this will allow the hoe crew to move more quickly through the field.

Applications of herbicides are also an option. When considering herbicides, growers must first determine if they have necessary equipment to apply the herbicide correctly. It may be to the grower's advantage to have the herbicide custom-applied rather than go to the expense of building a boom sprayer, which can apply broadcast applications, as well as directed or shielded applications. When combined with good cultural and mechanical practices, herbicides offer control for many weed species. The choice of herbicide depends upon the weed species, application timing and the grower's cultural practices.

Herbicide can be applied to peppers in several ways. It can be applied prior to planting to control any emerged weeds. It can be applied before the pepper seed is planted and mechanically incorporated into the soil. This method is called preplant incorporated. The herbicides used in the preplant incorporated method affect germinating weed seeds and cannot be water-incorporated effectively. In addition, a pre-emergence application can be applied to the soil surface following seed planting and incorporated through irrigation. These herbicides affect the germinaton of weed seeds and will not control emerged weeds.

A postemergence herbicide application will control emerged weeds. The weeds absorb through their leaves and stems and translocate the herbicide to the site of action. It is important to apply these herbicides to actively growing weeds. Weeds stressed due to environmental conditions will not absorb and translocate the applied herbicide as effectively. The use of adjuvants may be required with these herbicides to improve absorption and retention on the leaf surface.

A more complicated method is a 'shielded' application. The herbicide is sprayed on the emerged weeds, but the pepper crop must be shielded from the herbicide spray. Another complicated method is post-directed spraying. The herbicide is applied postemergence to the weeds with the application 'directed' to the soil at the base of the pepper plant. Such applications require shields to protect the pepper plants. Herbicides used in this way may or may not require weed-free soil at the time of application. For those requiring the area to be weed free, the application is usually made following cultivation. Such herbicides must also be incorporated mechanically or with irrigation water for proper placement of the herbicide. Herbicides are effective additions to pepper production systems, but only someone knowledgeable about herbicide application should use them.

DISEASE AND PEST CONTROL

Diseases and pests are serious constraints on pepper production in the field. The most common and serious diseases and pests found on peppers in the field are discussed in detail in Chapter 10.

GREENHOUSE PRODUCTION

A greenhouse represents the ultimate climate modification for pepper production. It protects peppers from adverse climate and pests and provides an elevated temperature year round. The grower has the opportunity to control temperature, humidity and even day-length. A greenhouse is covered with a transparent/translucent material so that sunlight can enter (see Fig. 7.7). The absorbed solar energy is converted to heat which elevates the greenhouse air temperature.

A greenhouse is a building that contains a production system where various operations relating to the propagation, growing and harvesting of plant material take place. Growing peppers in a greenhouse is similar to growing tomatoes, a crop where extensive greenhouse research has been done. When comparing peppers with tomatoes an important difference is the relationship between growth and fruiting. According to greenhouse managers tomato vegetative growth and fruiting are inversely related. To get good flowering and fruiting, the vegetative growth must be kept rigidly under control. With peppers, there is a direct relationship between growth and fruit development. Peppers need strong growth to produce early and prolific fruit development. Greenhouse production of peppers requires relatively high inputs

Fig. 7.7. Pepper seedling production in a greenhouse.

of nutrients and energy for optimal control of growth and product quality. Pest control is also needed to produce a quality crop.

Production Areas

Greenhouse pepper production has traditionally been located near population centres. The Netherlands is acknowledged as the world leader in intensive greenhouse pepper production (Buitelaar, 1989; Welles, 1992). Improved transportation has allowed peppers grown in The Netherlands to be sold in the United States. In 1995, The Netherlands had approximately 1100 ha of sweet pepper production in greenhouses. The most concentrated area for green-houses is a triangular area that includes Rotterdam, Utrecht and Amsterdam. Bell peppers are The Netherlands' top volume item from greenhouses. In 1995, in excess of 180 million kg of bell peppers were produced, with 18 million kg exported to the USA. About 60% of the export to the USA is of 'coloured' bells. The colours most often shipped are yellow and red with purple (lilac) and orange being shipped in smaller volume. The 1995 acreages for coloured blocky peppers in The Netherlands were 437 ha of red, 285 ha of green, 224 ha of yellow, 31 ha of orange and less than 25 ha for lilac. Other countries also produce peppers in greenhouses. In the United Kingdom, peppers are grown in greenhouses or polythene structures on about 70 ha (Fletcher, 1992). Plastic greenhouses have also expanded in the mild winter climate areas of the world. In 1987 the plastic greenhouse industry of the Almeria province located in the south-eastern part of Spain expanded to more than 13,000 ha (Castilla *et al.*, 1989).

Site Location

The ideal location for a greenhouse is where it will receive high winter light intensity, moderate winter temperatures, low humidity and easy access to markets or transportation points. The easy availability of existing utilities helps reduce establishment costs and will affect fuel costs. Because sunlight is the major heating source, avoid places where trees or buildings may shade the greenhouse. Windbreaks, if planted in a proper place, can reduce heating costs.

Construction

When considering greenhouse designs for pepper production, three major factors should be considered: load limitations, light penetration and cost. The primary load considerations include snow and wind. Roof slopes of at least 28° and heated air in the greenhouse should prevent snow accumulation on the roof. Bracing along sides of the greenhouse and roof should be sufficient to withstand wind. A concrete footing is preferred for a permanent greenhouse. A wide door at one end of the greenhouse will ensure easy access for equipment. Without sacrificing strength, support structures should be kept to a minimum to maximize light penetration. Glazing materials should be highly transparent. Overhead electrical lines, irrigation systems and heating ducts should be kept to a minimum. Support structures should be painted with a reflective, light-coloured material for maximum light reflection.

Greenhouses are also known as glasshouses because glass used to be the standard covering material. However, plastic-glazed greenhouses have several advantages over glass greenhouses, the main one being cost. Plastic is also adapted to various greenhouse designs, generally resistant to breakage, light-weight and relatively easy to apply. There are five major types of plastic coverings: acrylic, polycarbonate, fibreglass reinforced polyester, polyethylene film and polyvinyl chloride film.

Acrylic is resistant to weathering and breakage and is very transparent. Its ultraviolet radiation absorption rate is higher than glass. Double-layer acrylic transmits about 83% of light and reduces heat loss by 20–40% over single-layer. This material does not yellow. Its disadvantages are that it is flammable, very expensive and easily scratched.

Polycarbonate resists impact better and is more flexible, thinner and less expensive than acrylic. Double-layer polycarbonate transmits about 75–80% of light and reduces heat loss by 40% over single-layer. This material scratches easily, has a high expansion/contraction rate, and starts turning yellow and losing transparency within a year (although new varieties with UV inhibitors do not yellow as quickly).

Fibreglass reinforced polyester (FRP) panels are durable, attractive and moderately priced. Compared with glass, FRP panels are more resistant to impact yet transmit slightly less light, and transmit less light as they weather. This plastic is easy to cut and comes in corrugated or flat panels. It provides superior weatherability only when coated with Teflar™. Fibreglass has a high expansion/contraction rate.

Polyethylene film is inexpensive but temporary, less attractive and requires more maintenance than other plastics. It is easily destroyed by ultraviolet radiation (UV) from the sun, although film treated with UV inhibitors will last 12–24 months longer than untreated film. Because it comes in wider sheets it requires fewer structural framing members for support, resulting in greater light transmission. Using a double layer of 6 mm polyethylene on the outside and 2 mm as an inner barrier will help to conserve heat; this inner layer will also help to reduce water condensation. The inner layer should be 2.5–10.0 cm from the outside layer with layers kept separated by a small fan (creating an insulating dead air space) or wood spacers. The two layers of polyethylene film reduce heat loss by 30–40% and transmit 75–87% of available light when new.

Polyvinyl chloride film has very high emissivity for long-wave radiation, which creates slightly higher air temperatures in the greenhouse at night. UV inhibitors can increase the life of the film. It is more expensive than polyethylene film and tends to accumulate dirt, which must be washed off in winter for better light transmission.

Fig. 7.8. Peppers grown directly in the soil in a greenhouse.

Soil

Greenhouses may use native soil for pepper production (see Fig. 7.8). If native soil is used, the greenhouse should be constructed on level sites with deep, well-drained soils such as a sandy loam. A source of good quality water is also important. High salt concentrations in either the soil or the water can significantly reduce yields. If native soil is used a soil test should be taken before planting to determine the amount of fertilizer to apply for each crop. All phosphorus and potassium fertilizers should be applied before planting and incorporated directly into the soil. Nitrogen fertilizers should be applied in split applications, part before planting and the rest as needed during the growing season. Nitrogen fertilizers can be applied as side-dressings or through a drip irrigation system. Secondary and minor fertilizer elements should be applied as needed. Methyl bromide has been used to sterilize the greenhouse soil, but with the possibility of methyl bromide being banned, alternatives are needed. Steam sterilization of soils is possible, but highly expensive. Therefore, growers have switched to soilless medium.

Hydroponic Culture

Hydroponic culture of greenhouse vegetables involves the production of crops in sand, gravel or artificial soilless mixes in bags, tubes, tubs, tanks or troughs designed to allow the circulation of nutrient media needed for crop growth (see Fig. 7.9). Unlike conventional soil culture, hydroponic culture of greenhouse peppers is less forgiving and requires intense management. Although present automation systems can minimize fertilization and irrigation labour inputs, continuous monitoring of the system is important. Growers must be highly knowledgeable about plant growth, nutrient balances, cultural media characteristics and plant physiology.

It is estimated that 80% of greenhouse peppers are grown in soilless material and the industry is moving towards 100% by the end of the decade. The Netherlands grow 100% of their peppers in rockwool (see Fig. 7.10). The basic system used in The Netherlands consists of a container through which the nutrient solution is applied by drippers, typically one per plant. The fertilizer is usually supplied by an automatic controller. Irrigation timing may be manual or automatic. The most common medium is horticultural rockwool which is mainly used in the rigid slab form. The pepper seedlings are usually started and grown in a rockwool cube. Horticultural rockwool is a specialized product made in an insulation factory by melting volcanic rock in a blast furnace and spinning in into fibres. These are bonded together to form light rigid slabs of growing medium. Properties of the material are that it is sterile, light and inert. Most importantly, it has void space of 97%. This gives it the ability to hold very large volumes of water, while still retaining adequate air. The standard

Fig. 7.9. Pepper roots in a hydroponic system.

rockwool slabs used are 75 mm thick. The wrapping is slit at its base and the
slab allowed to drain freely. Although there are some differences between
individual rockwool products, this typically gives about 22% air and 75%
water by volume in the slab. Because of the increase in aeration with height,
when a cube is placed on the slab, the aeration at the top of the cube rises to over
40%. This is an important factor in avoiding crown rot. The typical volume of
rockwool currently used is about 1.4 l m^{-2} of greenhouse area.

Temperature Control

Many factors can affect the growth of pepper plants. The most important for the
grower to monitor is air temperature, especially at night. Heating normally
constitutes the major energy requirement for a greenhouse. Regulating air
temperature in the greenhouse is important for both vegetative growth and
fruit production. Greenhouse cooling is also important. Evaporative cooling is
the most efficient and economical way to reduce greenhouse temperatures in
areas with low humidity. Proper ventilation is also important, not only for
temperature control, but also to replenish carbon dioxide and control relative
humidity inside the greenhouse. Relative humidities of about 90% will

Fig. 7.10. Peppers grown with rockwool in a greenhouse.

encourage disease problems. Roof ventilators are seldom used on plastic houses, which instead use side vents to provide both ventilation and cooling. Vents should be installed as high on the wall as possible. Roof shading may be required in the late spring or early autumn if daytime temperatures become too high. Various shading materials that can be sprayed or brushed on are available from greenhouse supply companies. However, shade compounds must be removed when cool weather sets in. Shade cloths with various degrees of shading are also available.

    Heating, cooling and ventilation should be automated to save labour and to ensure proper temperature control. Polyethylene ventilation tubes with perforation holes sized about 7 cm along the tube suspended from the peak of the house from one end to the other help to mix cooler air evenly with warmer air, preventing draughts.

Light

Lighting is as important as temperature. Plants need sufficient light to maintain healthy growth and fruit setting throughout the production season. Cultivars have been developed specifically for greenhouse production. These cultivars do very well under limited photosynthetically active radiation (PAR). Peppers grow best in light whose wavelengths range from 400 to 700 nm. Most greenhouse coverings will accommodate these short waves of visible light. Polyethylene and fibreglass tend to scatter light, while acrylic and polycarbonate tend to allow radiation to pass through directly. Scattered or diffuse light tends to benefit plants by reducing excess light on upper leaves and increasing reflected light to lower leaves.

The recommended light level for pepper seedlings is approximately 35 PAR for 18 h day^{-1}. Demers *et al.* (1991) studied supplemental lighting on young pepper plants. They found that supplemental lighting substantially increased the specific weight before and after drying, the percentage of dry matter and the early, commercial and total yields, the total number of fruits harvested and the average weight of commercial fruits. Young plants subjected to supplementary lighting treatments yielded 1–2 weeks earlier than the plants that received only natural lighting. Supplementary lighting of 125 μmol m^{-2} s^{-1} for 20 h was only marginally better than the same intensity for 16 h.

Carbon Dioxide (CO$_2$) Enhancement

The introduction of supplementary carbon dioxide into the greenhouse has been found to significantly increase the yields of greenhouse peppers (Portree, 1996). Supplementary carbon dioxide is most effective on days when the greenhouse has been shut up for several days with no ventilation. During the winter months it takes longer to grow a plant to the final plant out state. Developing a strong vegetative plant before fruit set is important. Fruit ripens about 7–9 weeks after fruit set. Carbon dioxide enrichment to at least 800 ppm is recommended for growing plants. Maximum results have been achieved by injecting 1000–1500 ppm CO$_2$ into the greenhouse using propane burners or other CO$_2$ generators. Normal concentration of carbon dioxide in the atmosphere is 330–350 ppm.

Integrated Pest Management

Integrated pest management (IPM) is a holistic approach to the management of pests. IPM does not exclude the use of pesticides in the greenhouse. Rather, pesticides are used in combination with cultural, natural, mechanical and biological control as well as insect monitoring to maximize the effectiveness of control methods (see Fig. 7.11). Reduced use of pesticides under more effective timing schedules reduces not only the adverse effects of these chemicals on the environment and people, but also the chance of pests developing resistance (CPI, 1997/1998).

Fig. 7.11. Ladybirds controlling pests on bell peppers.

Propagation and Growing

Pepper plants can be made to behave as perennials under greenhouse conditions. If circumstances justify, one planting can be made to last for several years. This form of management saves the cost of the crop changeover (sterilizations, seed cost, propagation costs), but this gain must be weighed against the fact that the peppers will occupy the greenhouse in periods of low returns and prevent the growing of a more profitable crop. The decision whether to keep the plants longer is also greatly influenced by the health of the plants and it is certainly not viable to carry on a crop in which plants have died.

It was reported that in several greenhouse trials peppers grown on rockwool with white reflective mulch provided better plant growth and yields than did red light-reflecting mulch. Seeds are germinated in rockwool plugs 25 mm × 30 mm. The plugs are wetted with an electrical conductivity (EC) of 0.5 mS cm^{-1} and a pH of 5–6. Soluble salts can build up in the rockwool or growth medium. The build-up of soluble salts increases the likelihood of blossom-end rot. The liquid fertilizer temperature should be near 26°C. The media temperature should be maintained at 26°C until the plants emerge, then reduced to 24°C for the day and night temperature. To reduce the rate at which plugs dry, maintain a humidity level of 60–80% in the greenhouse. The plugs must remain moistened to at least 70% of their saturated weight during the germination process.

The seedlings in the plug are transferred into 75–100 mm rockwool blocks when the first true leaves appear. Some growers will invert the seedling. This process shortens the stem. Inverting may slow plant growth by 3–4 days, and the final plant is shorter and less likely to fall over when handled. The blocks should be wetted with a 2.5 EC fertilizer solution prior to transplanting.

The quality of the transplant will determine the final yield. A poor transplant with no lower leaves and which is etiolated will not yield as well as a healthy transplant. A hardening period a week before the plants are moved from the propagation area to the greenhouse will help. At 30 days, space the plants at 20 m^{-2} and set the block temperature to 21°C. Transplants are usually 6 weeks old and weigh about 40 g. However, a 7–8-week-old plant can provide a better transplant with thicker leaves and more dry matter.

The young plants are watered with a complete fertilizer solution. The plants should not be overwatered. Rockwool blocks can be watered with a nutrient solution once they dry down to 70% of their saturated weight. A final plant density varies from 2 to 3.5 plants m^{-2}. Peppers need more room than tomatoes. Generally rows are kept 1 m apart. The minimum plant distance should be 60 cm in both directions.

Peppers are self-pollinating in a greenhouse. Studies have indicated that pollination with bumble bees or honey bees reduces the days from fruit set to harvest. The pollination aid may also increase the percentage of extra large and larger fruit, and decrease the amount of deformed fruit. Bumble bee effectiveness appears to be related to cultivars. For example, studies in Holland with bumble bees have shown that cultivars like 'Eagle' have a significant pollination response, whereas 'Mazurka' does not. Honey bees can be introduced 3–4 weeks prior to flower development to aid in acclimatization of the hive. The hive needs a food source during the introduction period.

Peppers are sensitive to sodium which can reduce yields and fruit weight. If the pH drops to 5 for prolonged periods of time a manganese toxicity may become apparent. The damage is seen as 'burn' spots on the leaves near the top of the plant. Warm areas of the greenhouse with high transpiration rates tend to show the damage first. Boron deficiency is expressed as yellow discoloration of growing tips 30 cm below the top of the plant. Veins of the leaves turn brown. This condition is highly visible when the leaves are held up to the light and results from poor root growth, because boron is taken up by the young root tips.

Young transplants naturally branch into two or occasionally three shoots. This occurs after the fifth or eighth node. Plants are normally trained to two stems each. Because of their brittleness peppers need adequate support. The support system most commonly used is string. Strings are tied to stems and supported from overhead wires 2.5–3.0 m above. At about 4 weeks after planting out, two of the strongest shoots are selected. Stems are tied to strings supported from the overhead wires. The stem requires twisting around the string stock wire every 10–14 days. Failure to maintain regular pruning can result in lost production and reduced plant growth rate. Pruning should be continued near the top 10–15 cm of the plant. Alternate rows should be pruned or trained to reduce potential changes to the climate caused by plant stress. The option exists to train three stems instead of two per plant.

Ideally a pepper plant will set a fruit for every two leaves. When the lateral branches have four leaf axils above the first fork, the flowers are allowed to set. Misshapen and diseased fruit should be removed as soon as possible. Side shoots are removed as soon as possible allowing for better light penetration and larger flower development. The option of leaving more leaves per shoot should be considered when light intensity is high. This will prevent sun scald of the fruit. Secondary flowers at the axils of these leaves are of poorer quality. It is important that the first flower produced by the young pepper plant is removed. If this is not done and a fruit sets, early extension growth is reduced and the plant will have a tight centre with a poorly developed fruit.

Under greenhouse conditions, it can take up to 11 weeks from fruit set to colour development. A target of 6–7 fruits m^{-2} per week is realistic for production. Production greater than this rate may hinder calcium uptake and will result in blossom-end rot. It is recommended that a blunt-ended knife is used to cut the fruits off the plants. This will ensure a clean cut preventing stem infection and the blunt end will not damage fruit nearby. Scissors are not recommended because the rough wound sites they leave may encourage *Fusarium* infection. Fruits are picked when they are 85% fully coloured. The fruits are picked one to three times a week.

Storage conditions for greenhouse peppers are similar to those recommended for field produced fruits. The fruits are stored at 7–8°C if coloured and 10°C if green, with a relative humidity of 90%. Peppers are sensitive to low temperature and low humidity, especially if the fruit is stored and then exposed to temperatures of 19–21°C (Lownds *et al.*, 1994). Storage temperature also affects the time it takes to colour in storage. For example at 70% colour, yellows take 23 days to reach maturity at 8°C and 7 days at 24°C whereas reds take 13 and 10 days respectively. Relative humidity needs to be high to prevent fruit desiccation. Pepper fruit will lose its firmness after 2% moisture loss or show shrivelling at 6% moisture loss.

Disease

Because peppers germinate and emerge slowly, they can be particularly susceptible to damping-off. Using seed treated with a fungicide will help prevent seedling losses. As the plants mature, early disease identification and plant removal is important. Using virus-resistant cultivars when available is helpful. Only transplant healthy seedlings and discard any that are weak. Plants that are diseased should be rogued early before routine maintenance begins.

Spraying the seedlings with a 10% skimmed milk powder containing at least 35% protein will control virus spread when handling seedlings in the greenhouse. When working with plants, workers should dip their hands in a skimmed milk solution to reduce the spread of viruses. If milk is inconvenient, spraying the hands with rubbing alcohol also works. In addition, good sanitation by cleaning shoes and tools will reduce the spread of disease. Foot baths of quaternary ammonia or another viricide product at the entrance of the greenhouse help to keep diseases from entering the greenhouse. Restrict visitor access to header walkways only and do not allow the crop to be handled by visitors. Remove all crop debris at the end of the cropping season from the greenhouse site. Viruses can survive in dry plant debris for as long as 25 years. If possible, after each season, pressure wash the entire greenhouse interior and all carts and tractors, especially the tyres, which are used in the greenhouse. To reduce infections by tobacco mosaic virus (TMV), never allow smoking in the greenhouse.

ROW TUNNELS

One of the most effective means of altering microclimates under field conditions is accomplished with row covers or tunnel planting systems. Row covers are less expensive than greenhouses, but provide the pepper plant with a modified growing environment. Row covers are flexible, transparent coverings that are installed over a single or multiple rows of peppers to enhance growth and yield.

O'Dell *et al.* (1979) studied the effect of plastic tunnels on early season bell pepper production. They constructed row tunnels using '36-inch-wide' (~1 m) wire reinforced plastic. The rows were laid out in an east to west direction so that the prevailing wind would provide some tunnel ventilation. Semicircular hoops of No. 9 wire were placed over each of three rows at 2 m intervals. Wire-reinforced plastic was stretched over the hoops using very short pieces of the No. 9 wire to peg the plastic to the soil. Ends were left uncovered but bales of hay were available to block the tunnel ends in the event of cold weather. Plastic tunnels were about 35 cm high and 30 cm wide at ground level. The tunnels withstood high winds and heavy rains. They were then removed after the danger of frost had passed. Fruit set was excellent and production began about 2.5 weeks ahead of uncovered plants transplanted at the same time or later in the season. The reinforced plastic and wire hoops were reusable for at least 5 years.

Dainello and Heineman (1987) tried a less expensive system than the row tunnel, by using a depressional planting technique. With this technique, the plants are placed in the bottom of a trench and covered with a polyethylene sheet. When compared with peppers transplanted to regular raised flat-topped beds, the slitted clear polyethylene covered trench resulted in a 14% increase in the number of pepper fruits being harvested in the first harvest. The treatment also increased total yield by more than $2000 \, \mathrm{kg \, ha^{-1}}$ over the standard method.

CONCLUSION

There are several other factors that must be considered before one can expect to grow a bountiful harvest of peppers. Disease, pests and abiotic disorders, such as air pollution, must all be taken into account when growing peppers. The economics of pepper production is not discussed in detail because of its complexity. However, if it is not profitable to grow peppers, there will be no need to produce them. Even with all the difficulties, it is reassuring that pepper production is continuing or increasing in so many areas of the world. For pepper production to be found from the humid tropics, to the dry deserts, to the cool temperate climates makes it a very versatile crop.

8

HARVESTING

INTRODUCTION

The fruit stage for harvest is dependent on the final use. Most fresh peppers are harvested at the physiologically immature (horticulturally mature) stage, while the dehydrated and mash industries want physiologically mature fruits. For the canning of green New Mexican type pepper, a processor demands that all pods are free of any red colouring. This is because as the pod matures the 'skin' begins to stick and peeling of the pod is hampered. In addition, if the fruits are to be shipped long distances, care must be taken to reduce the 'field heat', so that respiration is reduced and the fruit will be saleable at its destination. Therefore, yield is dependent not only on the growing environment and cultivar but also on the time of harvest.

HAND HARVESTING

The majority of peppers in the world are harvested by hand (see Fig. 8.1). The reason that the pepper industry as a whole is still based on hand harvesting is quality. Hand-picked peppers are of a higher quality because humans can instantaneously reject mouldy, underripe, overripe or damaged pods. Humans also pick fewer leaves and stems as they harvest. Not only is the quality of the product better with hand harvesting, but there is an increased yield per unit area. The increased yield is associated with less damage to the plants as human pickers move through the field. Machine-harvesters cause more damage to plants than human pickers. Machine-harvested plants also take longer to recover and set more fruit.

   Because quality and yield are increased with human pickers, they are still the first choice to use. However, affordable human labour is becoming scarce, and costs of hand harvesting can escalate to a point where human labour is too expensive. For example, in the USA the chiltepin type pepper is a 100%

Fig. 8.1.   Hand harvesting peppers.

imported product. Labour costs in the USA, to harvest the small chiltepin fruits by hand, would put the price of the fruits beyond that which consumers would pay.

HARVESTING AIDS

Some harvest machines do not actually remove the pods from the plants, but are referred to as 'harvesting aids'. These machines assist the human harvester during the harvest (see Fig. 8.2) operation. A standard type of harvest aid carries the human pickers through the field, instead of the picker walking. The picker rides in a chair and reaches down to pick the ripe fruits. Some harvest aids have conveyor belts that move the picked fruits from the human harvester to a centrally located container.

MACHINE HARVESTING

Because human labour may become cost prohibitive, considerable research has been conducted to develop a mechanical means of harvesting peppers.

Fig. 8.2. An example of a machine harvest aid for harvesting bell peppers.

Research programmes in several countries have tried to develop a machine to mechanically harvest peppers. Marshall (1997) has estimated that more than 200 mechanical pepper harvesters have been built by 75 different groups, including private inventors, processors, agricultural machine manufacturers and public research institutions. Some machines remove fruits from the plant while others cut the plant off at the ground, then the plant and fruits are conveyed up to a unit where fruit are removed from the plant frame. Mechanical harvesters reduce the human involvement of harvesting to a driver and a few individuals sorting the pods. The fruits are then conveyed to a container on a trailer for transportation to the processor. Usually, machine-harvested peppers are used in the processing industry, instead of the fresh market. The bruising of the fruits during harvest reduces the shelf-life of the fruits. The bruising is not a problem with processors because they will usually process harvested peppers within 48 h.

A number of different techniques have been tried for machine harvesting of peppers. In California, modified tomato harvesters were tested on bell peppers that were destined for dehydration. Through various design changes, the tomato harvester's original shaker bed was replaced with a very aggressive, counter-rotating rubber roll cleaning bed. The plant stem was cut at the ground and the plant was carried up to the cleaning bed where a few individuals moved the plants around with hoes to increase exposure to the rollers. Also in California, a harvester with rigid, plastic fingers protruding from facing belts to remove New Mexican type peppers was tested (Lenker and Nascimento, 1982). Neither harvester was used after the initial development stage. In New Mexico, a machine with an inclined, counter-rotating brush and flap units similar to those used in a cotton stripper machine was used on mature New Mexican red pods. The pods were for processing at a dehydration facility. The machine seemed to work satisfactorily, but human labour was available, and thus the machine's use was discontinued.

The University of Georgia tried to design a machine to harvest pimiento peppers for processing (Fullilove and Futral, 1972). The machine had rigid fingers combing horizontally and vertically through pimiento plants. This machine was abandoned because it uprooted plants and had unacceptable pod damage. The engineers followed up on their early design with a machine that used a twin, open-helical harvest head. The twin double open-helix design was satisfactory for harvesting the pimiento pod type. Modifications to this machine by other researchers included the vertically oriented, twin, single open-helix concept. Another model had three rotary brush and finger stripper units with vacuum- and pressure-assistance. Other modifications to the basic helical harvest head included single, double or triple open-helixes, a large horizontal open-helix and an inclined closed helix. There was an attempt to reduce the aggressiveness of the harvest head by fitting a cylindrical pipe inside the helixes. The cylindrical pipe's outside diameter equalled the inside diameter of the formed open-helix.

In Texas, Posselius and Valco (1985) tested the concept of zone harvesting. Zone harvesting is harvesting horizontal layers in the lower portion of the plant, and then harvesting the upper portion of a plant at a later harvest. Because peppers mature from the base of the plant up, a range of maturity exists in the plant canopy. They successfully tested the use of shortened helixes. For the first harvest they used helixes one-quarter, one-third, or half as long as the original welded helix assembly. The first harvester proved to be useful for harvesting the lower portion of the plant and then some weeks later with the original full-length helix attached to the harvester they harvested the remainder of the plant.

Marshall and Esch (1986) reported that pepper recovery generally increased with helix rotational speed. No consistent relationship was found between damage and helix speed. Recovery and damage were unrelated to harvester travel speed for yellow bell, hot banana, hot cherry or sweet cherry types in the range from 0.5 to 3.0 km h^{-1}. Harvesting at higher travel speeds may not be efficient because of the increased volume of product and trash, the limiting capacity of trash removal equipment, and the additional hand labour needed to sort the material. This could also limit the number of rows harvested per machine.

Marshall and Esch (1986) also found that the average recovery and damage were unrelated to in-row plant spacings of 76, 152, 305 and 457 mm or at 200 and 400 mm row widths. Pod damage was negligible in cherry and jalapeño pod types (see Fig. 8.3). The hot banana pod type had more damage, typically 5–8%, but several test plots had negligible damage. With bell pepper pods the harvest damage was excessive at 10–20%.

In addition to the helical concept, a comber design has been applied to pepper harvesting. Axial combers, which are parallel to the row, horizontal rotary finger comber-brushes with vacuum and pressure assistance, and inclined rotary bars with fingers combing in a rearward and upward movement, have also been examined. Other combers tried have been inclined drapers with upward combing fingers, horizontal axle with radial fingers combing upward, an inclined rotary cone with fingers, fingers combing vertically, and inclined fingers combing rearward and upward. A second basic comber design was the transverse combers, these have heads that are transverse to the row of pepper plants.

A harvester with an inclined, axial, counter-rotating element with four parallel round bars was used successfully to harvest chile piquins (Bosland and Iglesias, 1992) (see Fig. 8.4). The majority of the chile piquin crop production in Mexico is a local endeavour with villagers collecting the fruits from wild

Fig. 8.3. A mechanical harvester picking jalapeños.

Fig. 8.4. An experimental mechanical harvester picking chile piquins.

plants. In the United States, all chile piquins are imported. The majority of the crop is imported from Mexico where the plants grow wild in the mountains. Chile piquins are not grown commercially in the United States because of high costs associated with hand harvesting the small fruits (1–2 cm in length). A piquin-type chile pepper, 'NuMex Bailey Piquin', was developed to be harvested by the machine. The plants were harvested when approximately 85% of the fruits were mature red without prior treatment with a fruit-ripening agent. The plant had the deciduous fruit trait which allowed the fruits to be shaken from the plant by the harvester. The harvester shakes the plant and an attached conveyor belt carries the fruits to the rear of the machine for collection. This procedure is analogous to that used on some nut trees, for example pecans.

Other harvesting designs consist of an axial universal chain with perpendicular fingers engaging the plant base and with outer tips supported and moving up an incline ramp. In addition, some other designs incorporated a transverse sickle bar cut-off, inclined parallel stripping bars with a front end hinged and rear end moving up and down via a crank throw. The pepper plant stems are cut off with the sickle bar, then two spring-loaded V-belts invert the plants with a subsequent downward-stripping force using a cylinder with finger beaters.

A forage chopper has been used in Oklahoma for Bahamian peppers. An interesting and somewhat different approach was the use of high-pressure rotating water spray to harvest tabasco peppers. Many other designs have been tried but Marshall (1981), after evaluating many harvesting concepts, found the twin, double open-helix the most acceptable concept for harvesting most major pod types of peppers grown commercially in the United States.

PLANT HABIT

Plant traits that aid in machine harvesting are a uniform and concentrated fruit set. A plant frame that allows the fruit to set somewhat high on the plant, so the harvest heads can move through the plant easily, is also important. Sundstrom *et al.* (1984) found that a high nitrogen rate and high plant populations not only produced a favourable plant structure for machine harvesting, but also increased machine-harvested red tabasco pepper yields.

Very few peppers are being harvested mechanically today in spite of the extensive research done on pepper harvesters. A wide variety of principles have been evaluated by many researchers in North America, Europe and the Middle East. With increased cost of labour, renewed interest in harvest mechanization may result in improved harvesters. To machine harvest peppers successfully, cultivar development will have to occur concurrently with the development of the machines (Marshall, 1997). Cultivars specifically bred for machine harvesting will improve the efficiency of the machines.

TRASH REMOVAL

Mechanically harvested peppers will contain more leaves, plant branches, damaged fruits and other extraneous matter than hand-harvested peppers. Mechanical harvesters pick a greater percentage of misshapen or injured fruit when compared with the hand-picked harvest methods. Trash must be removed before the peppers can be processed. The increased trash problem with mechanical harvesting should be handled mechanically so that labour and harvesting cost are minimized. Several pieces of equipment have been tested for the auxiliary removal of plant trash. These have included a shaker bed trash removal system, a rotary rubber finger bed, a stationary sorting conveyor, and a counter-rotating smooth steel roll turning against a steel roll wound with a helix.

The trash removal efficiency and the amount of damage of four trash removal systems were tested by Esch and Marshall (1987) on the fruits of cherry and yellow wax type peppers. The systems tested were counter-rotating rolls (two sizes), combing belts and rubber starwheels. The effectiveness of counter-rotating roll beds was significantly enhanced by first feeding the harvested peppers through differential-speed, double, pegged combing belts. Esch and Marshall (1987) found no significant differences in pepper fruit damage among the five methods of trash removal tested. They tried a combination of one smooth and one helically wound (50.8 mm diameter, 50.8 mm pitch) rubber roller. This combination proved to be the most successful in eliminating trash with minimal damage to the peppers.

A cleaning unit with unidirectional rotating starwheels has been used in Israel to remove light trash (leaves, small branches and undersize fruit) from field-dried red New Mexican type peppers. During field tests in Michigan with yellow wax and cherry types, the machine performed poorly in removing trash (Esch and Marshall, 1987). They did conclude that the unidirectional rotating starwheels may have potential as an in-the-field size grader for processing-type peppers.

As development of machine harvesters for pepper continues, instrumentation and machines for automated removal of harvest trash, grading and sorting of peppers are also being developed.

SORTING AND GRADING

Sorting and grading peppers by machine would be labour-saving and more economical than hand labour. In fresh market packing operations, peppers are initially sorted according to colour and damage (see Fig. 8.5). Acceptable peppers are then packed for shipping into one of four to five classes according to size and shape. Machine vision offers the potential to automate many manual grading practices. As microprocessor speeds continue to increase and costs

Fig. 8.5.   Bell peppers being sorted before shipping.

decrease, machine vision will certainly become a cost-effective solution. USDA grading standards mandate that the entire pepper surface be inspected. Variations in size, shape and symmetry make it nearly impossible to mechanically manipulate the pepper into any standard orientation. Shearer and Payne (1990) concluded that six orthogonal views would be needed to adequately characterize surface colour and damage for sorting.

Colour sorting of peppers uses criteria which range from the pepper being too light a shade of green to rejecting peppers exhibiting even the slightest hint of red colour. Pale green or light-coloured peppers can often be attributed to cultivation practice. Colour variations from a slight reddish cast to bright red are a direct result of senescence. In either case, simple characterization of the visual spectrum of light reflected from the pepper surface should provide sufficient information to support an accept/reject decision. Shearer and Payne (1990) report that machine vision, when applied to the task of grading bell peppers for colour had accuracies of up to 96%.

MECHANICAL INJURY

Pepper injuries include scars, sunburn, disease infection, hail damage or even damage from mechanical harvesting. During machine harvest and postharvest handling operation, peppers undergo several transfers, each has the potential for causing mechanical injury to the pepper fruit. Mechanical injuries include abrasions, cuts, punctures and bruises, all of which can lower the market grade of the peppers, and reduce the subsequent shipping life of the pepper fruits. Again, Shearer and Payne (1990) hypothesized that by characterizing the colour of light reflected from the pepper surface, sufficient information could be obtained to support accept/reject decisions with regard to these damaged areas. They reported that detection of damage was more difficult than pod colour, and the highest accuracy they could obtain was 63%.

Wolfe and Sandler (1985) reported on the development of a stem detection algorithm using digital image analysis. The algorithms are the commands or parameters that 'tell' the machine sorter whether the item in its view is acceptable or unacceptable. The algorithm routine relied on analysis of angle patterns in the boundary chain code of profile images. When the algorithm was tested on cherry peppers, the performance was very good with error rates only in the range of 1.5%. Wolfe and Swaminathan (1986) used circular and linear Hough transforms for detection of stem and blossom ends of bell peppers. These locations were then used to determine the orientation of the peppers on a sorter with average errors of 8.1° between the measured and calculated orientation angles. Axial paired gradients and medial axis variance were then used to characterize bell pepper shape.

Marshall and Brook (1997) measured the impacts occurring in a pepper field and on a packing line with an instrumented sphere. They found bell peppers bruised mostly on their shoulders. Most bruising on packing lines occurred at transfer points between different pieces of equipment when the peppers fell or were propelled from conveyors on to uncushioned metal plates or rollers. The main problems on packing lines were the excessive height differences between line components, lack of control of rolling velocity and lack of cushioning on hard surfaces.

Pepper fruit detachment force is mainly a function of the gene system that involves pedicel characters in the fruit genotypes. Pepper fruits do not have an abscission layer in pedicels as do tomatoes. Marshall (1981) found that, in machine harvesting cultivars of diverse fruit characteristics, serrano fruits were the easiest to remove because of the small diameter of the pedicel scar at the point of attachment to the stem. In removing pepper fruit, the pull of force at the pedicel attachment site is correlated to the stem scar diameter, with pedicel attachment being very strong in cultivars with large fruit. Werner and Honma (1980) reported that easy fruit removal was positively correlated with fruit length, diameter and weight and that fruit detachment force was an inheritable character. Setiamihardja and Knavel (1990) suggest that if plant breeders want to select for low fruit detachment force, they should select for long, narrow, pendant fruits.

RED PEPPER HARVEST

One important horticultural aspect of machine harvesting of pepper is the timing of the harvest. Once-over mechanical harvest of red pepper (or paprika) which is based on a single destructive harvest may be the most efficient and simple method. Allowing the fruit to dry naturally on the plant before harvest can lower transport and storage volume and reduce the energy required to dry the fruit artificially in the dehydration facility. However, if fruits stay on the plant too long after red maturity, a loss in yields is possible. Cotter and Dickerson (1984) found that in New Mexico the yield of mature red chile fruit peaked in late October or early November (before first frost) and then declined through January. A significant red colour loss to the crop was also noticed as the harvest date became later. While a once-over mechanical harvest of mature red pepper or paprika early in the season could maximize yields, the result is a mixture of red mature and green immature fruit. Immature fruit will reduce the value of the crop by diluting the intensity of the red pigment in the processed product. It would be best to have all the fruits mature red at the same time.

ETHEPHON

One possible approach to reducing the number of immature fruits is to use a chemical to speed up the red colouring of the pods. Lockwood and Vines (1972) reported that ethylene gas was not effective in accelerating the degreening of pimiento peppers, but that ethephon (2-chloroethyl phosphoric acid, Rhône-Poulenc Co.) did significantly promote the degreening process.

The action of ethephon on ripening of peppers is affected by several factors. These include pepper type, cultivar, concentration of application (Batal and Granberry, 1982; Knavel and Kemp, 1983), number of applications (Cantliffe and Goodwin, 1975), air temperature (Knavel and Kemp, 1983) and crop maturity (Batal and Granberry, 1982). Ethephon has been successfully used to concentrate red fruit maturity (Cantliffe and Goodwin, 1975). However, ethephon has given variable results as a fruit-ripening agent on pepper. In many cases, leaf defoliation and fruit abscission occurred. This can offset the beneficial effects of ethephon on fruit ripening by reducing yield and quality. Bell pepper flower buds are known to abscise in response to ethephon (Tripp and Wien, 1989). Kahn *et al.* (1997) suggested a single application of ethephon at 2000–3000 µl l^{-1} as a controlled abscission agent to increase the percentage of harvested red fruit, while minimizing an excessive flower drop. This concentration increased the marketable fruits as a percentage of total harvested fruit mass, but decreased the total dry mass of harvested fruit. Ethephon application at the rate of 1500–3000 ppm induced defoliation and fruit abscission in pimiento and paprika, especially at later stages of fruit development (Batal and Granberry, 1982). They suggested that field applications of ethephon can be part of the overall production practices in the pimiento and paprika industry. Ethephon accelerated fruit ripening when applied to plants at stages closer to normal fruit maturity. Ethephon also increased fruit abscission when applied at later stages of fruit development. This concentrated ripening resulted in higher yields of usable fruit with improved quality for once-over harvesting of pimiento and paprika. Removal of green or immature fruit prior to harvest would certainly improve efficiency of mechanical harvesting.

Cantliffe and Goodwin (1975) demonstrated that a high single concentration of ethephon sprayed on pepper plants could cause more chlorosis, defoliation and fruit abscissions than repeated application at lower concentrations, and total yields can be substantially reduced. Cantliffe and Goodwin (1975) recommended a concentration of 100–200 ppm ethephon applied three times to provide a greater margin of safety than a single spraying of a high concentration. Ethephon concentrated maturity for once-over harvest without reducing the average fruit size nor was there an increase in the amount of spoiled fruits.

The effect of temperature on ethephon-induced fruit ripening in pepper has been of concern. High temperatures after ethephon treatment have accelerated fruit ripening, defoliation and abscission, while low temperatures have reduced or negated the effects of ethephon. Multiple applications using lower ethephon concentrations may offset these inconsistences.

An ideal harvest time is dependent on the pepper type grown. For example, a mature green pod of the New Mexican type that will be used for processing or fresh market feels firm when squeezed and is flat (has two cells), smooth, thick-fleshed, bluntly pointed and about 17 cm long, while a pepper harvested for paprika, will be semi-dried on the plant, disease- and blemish-free, and high in red colour content.

Multiple picks of peppers is common for most types. Even those picked at the mature red stage, i.e. red chile and paprika, could be picked several times because of the sequential setting and ripening of fruits.

Defoliants or desiccants, such as sodium chlorate, are often used to both accelerate fruit drying during wet weather, and aid in harvesting. Ethephon as a ripening enhancer may defoliate, as well as hasten maturity. This chemical will also increase the colour of red peppers that are harvested before frosts.

9

POSTHARVEST HANDLING

The postharvest handling of peppers is as crucial as the growing of the crop. Whether the pepper is used as a fresh commodity or processed, appropriate postharvest handling is essential for a quality product. The postharvest handling of pepper has several aspects. Eight major areas deal with the postharvest handling of peppers. Peppers can be fresh, canned, brined/pickled, frozen, fermented, dehydrated and extracted for oleoresin.

Fig. 9.1. Peppers being sold in a Guatemalan marketplace.

147

Whether peppers are intended for the fresh market (see Fig. 9.1) or for processing, quality should be maintained through all phases from field production to consumption. High-quality pepper begins with the selection of the proper variety and the purchase of quality seed. Before postharvest handling occurs, good cultural practices such as fertilization, irrigation and disease management in the field must be maintained to produce a high-quality crop. The level of stress that the crop endures in the field will influence yield, pungency, fruit colour and diseases. In general, peppers harvested from poorly managed fields will have inferior postharvest handling quality (Wall and Bosland, 1993).

FRESH

All peppers can be used fresh. Nevertheless, most of the pepper types that are used fresh have a thick, succulent wall such as bell, pimiento, New Mexican green chile, and jalapeño. Raw, fresh pepper fruits are among the richest known plant sources of vitamin C. The US Food and Drug Administration nutrient content descriptors for fresh peppers include 'fat-free, saturated-fat-free, very low sodium, cholesterol-free, low in calories, high in vitamin A, and high in vitamin C'. A good quality, mature, fresh green pepper is firm, bright in appearance, thick-fleshed and with a fresh, green calyx. Immature peppers are usually soft, pliable, thin-fleshed and pale green in colour. High-quality pepper is free from bruises and abrasions, bacterial, fungal and viral diseases, blossom-end rot and sunscald. Pepper harvested during rainy periods and shipped wet will probably develop diseases quickly, especially if not refrigerated.

Pod firmness or the degree of dehydration is the primary indication of pepper freshness (Lownds *et al.*, 1993). Fresh green pepper loses water very quickly after harvest and begins to shrivel and change colour within a few days if unrefrigerated. If stems remain they should be firm and green. Darkening, shrivelling or rotting of stems indicates that the pepper was not harvested recently. To ensure a pepper fruit of high quality, the fruit must have quick and proper cooling. All pepper types, but especially New Mexican green pepper, are highly susceptible to water loss, sunscald and heat damage. These problems are likely to occur if peppers are allowed to sit for more than an hour in direct sunlight (Boyette *et al.*, 1990). Fresh pepper harvested in the summer-time can have a pulp temperature of 32°C or more. For these reasons, peppers should be harvested in the early morning, placed in the shade and cooled as soon as possible. If the peppers are not cooled within 1–2 h, they will begin to show signs of water loss and softening. Whenever possible, pepper should be precooled, transported in refrigerated trucks and kept cool before processing.

Temperatures higher than 21°C greatly accelerate ripening through respiration and ethylene production. Refrigeration extends the shelf-life of pepper by decreasing respiration, water loss, colour change and postharvest disease development. Preferred cooling methods for peppers are forced air cooling and room cooling (Boyette *et al.*, 1990). Room coolers can be partitioned into sections. Harvested peppers with high field heat should be kept separate from other parts of the cooler used for storing previously cooled produce. Room coolers can be modified into forced air coolers relatively quickly and inexpensively by adding extra fans and partitions (Boyette *et al.*, 1989). With forced air cooling, the fans pull cool air through the boxes or bins of produce. Forced air cooling is an active cooling process and is much faster at removing field heat than room cooling (Boyette *et al.*, 1989).

Most fresh peppers can be stored for 2–3 weeks if kept cool at 7–8°C. The optimum storage temperature is 7–10°C, with a relative humidity of 85–90%. They are highly sensitive to freezing injury and susceptible to chilling injury. Chilling injury occurs below 4°C. The symptoms of chilling injury are softening, pitting and a predisposition to decay. Freeze damage occurs at 0°C. Fruit ripening is greatly accelerated by the presence of the natural ripening hormone, ethylene. Some fruits and vegetables produce much larger quantities of ethylene, as compared with the amount produced by peppers. For that reason, peppers should never be stored or shipped with crops such as tomatoes, apples or melons. As peppers ripen in storage, they will produce an odour that can be absorbed by pineapples.

Fresh pepper is typically stored and shipped in waxed corrugated boxes (Boyette *et al.*, 1990). Modified atmosphere packaging which keeps the relative humidity high, but allows for gas exchange aids in shipping peppers (Lownds *et al.*, 1994). Currently, bell peppers are shipped in a variety of containers around the world. In Mexico, standard containers are a 35 lb 1¼ bushel; 30 lb cartons/crates. In the Netherlands, 28 lb bushel and 1⅑-bushel cartons/crates; 28 lb 3.56 dekalitre cartons; 25 lb cartons; 14–15 lb ½-bushel, and a 11 lb flat carton are common. The US grades for bell peppers are US fancy, US No.1, and US No.2 (Table 9.1). Packaging for jalapeños and yellow wax pod types is ½-bushel crates/cartons, ⅝-bushel crates/cartons. For other types of peppers it is 1⅑ bushel crate/carton. There are no US official federal grade standards for chile peppers.

Table 9.1. Grades and sizes for bell peppers in the USA.

| Grade | Minimum size | | Uniformity of colour[a] (%) | Firmness | Shape | Damage[b] |
	Length (cm)	Diameter (cm)				
US Fancy	9	7.5	90	Firm	Excellent	None
US No. 1	6.5	6.5	90	Firm	Good	None
US No. 2	None	None	90	Firm	Fair	Light

[a]All green, all red or mixed are acceptable colour grades.
[b]Damage includes sunscald, freezing injury, decay, scars, hail, sunburn, disease, insect and mechanical.

CANNED

The most common pepper types to be canned are the New Mexican type, the pimiento and the jalapeño. An ideal, mature green pod of the New Mexican type suitable for canning is firm, flat (has two locules), smooth, thick-fleshed, bluntly pointed and about 15 cm long and 4 cm wide at the shoulders (Bosland, 1992) (see Fig. 9.2). The introduction of the 'Perfection Pimiento' and the invention of canning machinery for roasting to remove the skin began the pimiento canning industry in the USA. Pimientos intended for processing should be fully mature, bright red, thick-walled and firm. Weisenfelder *et al.* (1978) have described the ideal processing jalapeño as glossy, light green, 7 cm long, 2 cm wide, with an approximate capsaicin level of 1.6 mg 100 g^{-1} dry weight. Both New Mexican green peppers and pimientos are peeled prior to canning. New Mexico green pepper fruits are either flame roasted or steam peeled, whereas pimientos are steamed, flame roasted or lye peeled (Johnson, 1977; Flora and Heaton, 1979).

Pepper fruits are normally processed at 100°C. However, long exposure time to this thermal treatment or to pressure-processing can soften the fruits excessively. According to the US Food and Drug Agency, a manufacturer must thermally process acidified foods sufficiently to destroy all vegetative cells of microorganisms of public health concern (FDA, 1979). A pH value of 4.6 is the upper limit for preventing toxin formation by *Clostridium botulinum* (McKee, 1998b). Acidification of the product reduces the pH below 4.6, decreases the thermal resistance of microorganisms, and allows the canner to reduce the thermal exposure time and to process peppers at atmospheric pressures (Powers *et al.*, 1950). The pH of peppers has been shown to vary among early to

Fig. 9.2.   New Mexican green pods being washed before canning.

late season harvests and with degree of ripeness (Powers *et al.*, 1950; Powers *et al.*, 1961; Flora *et al.*, 1978; Flora and Heaton, 1979; Sapers *et al.*, 1980). Therefore, concentrations of acidulant must be adjusted to account for the pH variability of the raw fruits. Properly canned pepper fruits generally have a maximum shelf-life of 2 years.

Much research has been reported on the acidification of canned pimientos (Powers *et al.*, 1950; Powers *et al.*, 1961; Supran *et al.*, 1966; Flora *et al.*, 1978; Flora and Heaton, 1979). Citric acid is the most common acid used to acidify pimientos, although fumaric acid is also effective and less is needed to achieve the same pH level (Powers *et al.*, 1950; Flora and Heaton, 1979). The pH of processed pimientos can increase during storage. In a study by Flora and Heaton (1979), pH changed from an initial 4.37 to 4.59 after 12 months storage. The initial pH of the canned product should be sufficiently low to compensate for such a pH increase during storage, although acidification below pH 4.3 may create an undesirable flavour (Supran *et al.*, 1966). Peeling and processing techniques also affect acidification requirements. Lye peeled pimientos, with some NaOH residues, require more acidification than flame peeled pimientos, and blanched pimientos are acidified more rapidly than unblanched pimientos (Flora *et al.*, 1978).

In addition to pH considerations, fruit texture is a major concern for processors. Fruit softening of canned peppers can be minimized with calcium treatments. Calcium chloride is the preferred calcium source for both pimientos and jalapeños, although calcium lactate is also acceptable (Powers *et al.*, 1950; Powers *et al.*, 1961; Saldana and Meyer, 1981). In canned pimientos, the combination of citric acid and 0.02% calcium chloride with the product significantly increased both firmness and drained weight when compared with either treatment alone (Powers *et al.*, 1961). The firmness of canned jalapeños was increased twofold, without imparting bitterness, when 0.2% calcium chloride was added to the product (Saldana and Meyer, 1981). Concentrations higher than 0.2% did not further improve jalapeño firmness. Calcium hydroxide was unacceptable as a firming agent, because it raised the pH and precipitated in the can.

Jalapeños and a dried smoked form of jalapeños called chipotle are often canned in a brine called escabeche, which includes vinegar (acetic acid), oil, sugar and spices at an equilibrium pH of 3.7 (Saldana and Meyer, 1981). Changes in pungency may be a problem in canned jalapeños. Canned jalapeños, thermally processed at 100°C for 50 min contained higher capsaicin levels than fresh samples in a study by Huffman *et al.* (1978). However, in a later study, canned jalapeños actually lost one-half of the capsaicin level present in the raw jalapeños (Harrison and Harris, 1985). Those fruits were blanched for 3 min at 100°C, rinsed, packed in 93°C brine (2% acetic acid, 2% vegetable oil, 0.2% sodium chloride) and thermally processed at 100°C for 50 min. The capsaicin may have leached into the rinse water or the brine in that study.

BRINED/PICKLED

Peppers that are usually brined include cherry, wax, pepperoncini, jalapeño and serrano. The pickling or brining process involves adding sufficient quantities of salt and acetic acid to prevent microbial spoilage (Daeschel *et al.*, 1990). Peppers packed in this manner have superior textural qualities compared with canned peppers. Brining processes for pepper are similar to those for other brined vegetables such as cucumbers, okra, carrots and cauliflower. The primary factors in the quality and shelf-life of brined peppers are the initial field quality, the length of time between harvest and brining, the specific brine chemistry, the storage conditions and environment, and the addition of preservatives. The degree of mechanical injury before brining can affect the quality of the final product. Pepper fruits that are damaged and bruised during harvest and handling, or are subjected to sustained high temperatures after harvest soften easily. Also, mechanically de-stemmed peppers that are cut open are more prone to deterioration than hand de-stemmed peppers or whole peppers. Mechanical de-stemming, slicing or chopping should occur after an initial brining process for best quality.

The effectiveness of brining for preservation is related to the rate of acid diffusion into all parts of the fruits and the time required to reach an equilibrium pH of 4.6 or below. The primary areas for acid penetration are through the stems and calyxes and into the placentas (Daeschel *et al.*, 1990). The interior of fruit walls is the last area to become acidified, and the entire process can take at least 6 days. Therefore, the first week of brining is the most critical. Exposure to oxygen prior to brining can reduce the time for acid penetration to 1 day (Daeschel *et al.*, 1990). Blanching can also improve the rate of acid penetration into the fruit and therefore reduce the pH variability among fruit parts (Stroup *et al.*, 1985).

In most commercial operations, peppers are brined in a two step process. Fresh pepper fruits are placed in a primary brine to firm and to preserve the fruits. After a minimum period of 2–8 weeks (depending on variety and process), the fruits are removed from the first brine, washed, graded a second time, and then re-packed in a finishing brine. Only the best quality fruits are packed whole or sliced in the finishing brine. Vinegar and salt levels may be reduced in the finishing brine and spices may be added for flavouring. This second brine is usually added to the fruits in the final container, and the container is sealed and hot-packed. A large portion of fruits are never packed in the second brine, instead they are chopped and blended into other food products after the initial brining.

Brine procedures vary according to producer. In general, the initial brine has a sufficiently high salt and acetic acid concentration to ensure that the fruits retain rigidity and colour, and that microbial growth is prevented. The initial brine should have a maximum pH of 3.8, with 1–1.5% acetic acid by weight, and the solution should be saturated with food or pickle grade salt

(24–26% by weight). Salimeters or specific gravity meters can be used to ensure that the brine is at least 98% saturated. Final packing brines vary according to producer recipe, but are usually formulated for a pH of 4.2 or less.

Preservatives are often used in the initial brine to further prevent fruit softening and discoloration. Sodium bisulphite (0.5–1% by weight) is the most common preservative in pickled pepper. Sodium bisulphite is an effective preservative but imparts an off-flavour which should be leached in the second, packing brine. Bisulphite has also been implicated in certain food allergies experienced by asthmatics. Calcium chloride (0.25–0.50% by weight) can replace bisulphites, but is less effective at firming, can impart bitterness at higher concentrations, can darken fruit colour and requires additional salt for optimal rigidity of the fruit. Brined pepper fruits can be produced without preservatives by using a more costly refrigerated process. Cold temperatures allow the fruits to imbibe brine completely, while slowing the growth of bacteria which can soften the fruits. The first 4–8 weeks are important for fruit consistency in the brining operation; if the brining conditions are not optimal there is a chance that the fruit will become soft (mushy). After 6–8 weeks in cold storage, the fruits reach an osmotic equilibrium with the brine solution and further softening will not occur. Brined fruits without chemical preservatives then are packed in finishing brines with higher vinegar and salt concentrations than those with bisulphite.

Fruits can be stored in the initial brine for up to 9 months before packing, although most are held for 2–3 months. Most food companies in the United States do not put expiration dates on brined peppers.

FROZEN

One of the more modern ways to preserve peppers after harvest is to freeze them. Depending on the pod type harvested, preparation of the product before freezing may differ. New Mexican pod types are peeled before freezing, while jalapeños and bell peppers are not. Bell peppers and jalapeños are blanched before freezing. The peroxidase enzyme test is used to check if the blanching has been sufficient. The removal of the pepper fruit skin from the New Mexican pod type is a *de facto* form of blanching. If the peppers are kept at −35°C, the storage time for peppers may be up to 2 years. Preservatives are not used when freezing peppers. Another freezing method is the 'Individual Quick Freezing' (IQF). This is used for diced New Mexican, jalapeño and bell-type peppers.

The biggest loss of quality is attributed to the loss of green colour. A blanched product retains the green colour better than a non-blanched product. A non-blanched product has a shelf-life of 3–4 months, whereas a blanched product easily has a shelf-life of 12 months, the industry standard. Frequently, the product remains in acceptable condition for greater than 24 months at −32 to −35°C, but for marketing the product is labelled with a shelf-life of 12 months. For red pepper product, loss of colour does not seem to be a limiting factor for shelf-life. Unlike the green product, the red product maintains its colour without blanching. When jalapeños were blanched for 3 min, frozen and stored at −18°C, they retained only half of the capsaicinoids present in fresh jalapeños (Harrison and Harris, 1985).

FERMENTED

Hot pepper varieties of *C. annuum*, *C. frutescens* and *C. chinense* give bottled hot pepper sauces their characteristic flavours and pungency. The peppers are typically ground with 14–20% salt. Depending on the hot sauce recipe, the mash is either used immediately or aged for several months or years. During the process of ageing, fermentation occurs through microbial action and contributes to the unique aged flavour of the mash and, ultimately, the hot sauce (see Fig. 9.3).

Tabasco sauce®, produced by the McIlhenny Company of Louisiana, is probably the best known hot sauce. It is produced from fruits of *C. frutescens*. Hot tabasco pepper mash is aged for a minimum of 3 years in oak barrels before being used in Tabasco sauce® production.

DEHYDRATED

Large quantities of dehydrated pepper (see Fig. 9.4) are used in prepared meals, seasoning blends and in the canning industry. Dehydrated pepper types are New Mexican, cayenne, ancho, pasilla, mirasol, piquin and de arbol (Bosland,

Fig. 9.3.   Fermenting pepper mash.

Fig. 9.4. Red peppers being prepared for dehydration in a heated drying chamber.

1992). Dehydration of pepper for storage is an ancient art. The quality of red pepper and paprika products is based on pungency level, extractable red colour and flavour. The higher the number, the higher the amount of extractable red colour. Pepper must be processed and stored correctly to maintain high quality. Red pepper and paprika are dehydrated and sold as whole pods, or ground into flakes or powder.

As stated in Chapter 2, paprika is a very important world commodity. Any dehydrated, non-pungent, red pepper powder is 'paprika' in international trade (see Fig. 9.5). Besides the USA, an important producer of paprika is Spain. The major area for paprika production has been the Extremadura region along with León, Murcia and Catalonia. Fairchild (1938) wrote that in 1901 he found brilliant strings of red sweet peppers decorating the Spanish market-stalls. These peppers were used to make paprika and have a sharp, hot flavour and were dark red in colour and the shape and size of small apples. He found tons of paprika being produced. The three qualities of Spanish paprika are the 'extra', 'select' and 'ordinary'. The 'extra' quality is produced exclusively from pericarp tissue, i.e. the seeds, calyxes and pedicles are not ground with the pericarp. The 'select' grade is prepared from the pericarp and seeds. The seeds can contribute up to 10–15% of the total weight. The 'ordinary' is prepared by grounding pericarp, seeds, calyx and pedicel. The addition of other plant tissue besides the fruit wall lowers the colour, but increases the processed yield by as much as 25%.

Another historic and important production area is Hungary. Most of Hungary's paprika is produced in the regions around Kalocsa and Szeged. The spice paprika is exclusively grown outdoors, while the vegetable paprika is

Fig. 9.5. Dried pepper being milled into powder.

grown outdoors in the summer and in greenhouses in the cooler months. Hungary produces a wide range of paprikas from very mild to very hot. The Hungarian type paprika is also produced in Macedonia and Bulgaria. The fruit shape of Hungarian paprikas varies from tomato-shaped to triangular or heart-shaped to thin, elongated cones.

The Hungarian dried ground product is classified into eight types. Special Quality (Különleges) is the brightest red of all the Hungarian paprikas and has the mildest flavour with excellent aroma. Delicate (Csípmentes Csemege) is a very mild tasting product with colours that range from light red to dark mahogany. Exquisite Delicate (Csemegepaprika) is similar to Delicate but has more pungency, although at a very low level. Pungent Exquisite Delicate (Csípös Csemege, Pikant) is very much like Delicate and Exquisite Delicate in flavour and aroma, but is much more pungent. Noble Sweet (Édesmemes) has bright red colour with a mild pungency. Nobel Sweet is the class of paprika most often exported from Hungary. The red colour of the next three classes is reduced. Semi-sweet (Félédes) is a medium hot pungent product; Hot (Erös) is light brownish-yellow in colour and is the hottest of the paprika products. Rose (Rózsa) is pale red in colour but with a strong paprika aroma and is medium hot.

Traditionally, pepper was dehydrated by sun-drying. Originally, the fruits were spread on roofs or even on the ground, but damage by birds and rodents caused poor quality for processors. Thus, people began tying pepper together in strings (ristras) and hanging them along walls. This method was replaced by controlled artificial drying, now practised by virtually all commercial processors in the United States. Red colour retention mainly depends on the prevention of an oxidative process that reduces the original colour (Lease and Lease, 1956). Colour can fade rapidly if too much moisture is removed, but mould may grow if moisture content is high. While there is a market for whole dried peppers with good red colour, most of the pods are diced and dried. The diced fruits are dried to 4–6% moisture content. The pepper is then ground and rehydrated to 8–11% moisture, an optimal level for storage. Cold storage (3°C) is recommended (Lease and Lease, 1956).

Red colour retention is an important quality consideration for paprika and pepper powder and mainly depends on prevention of oxidative attack of the powder (Lease and Lease, 1956). Moisture content, storage temperature and atmosphere, light, harvest conditions and timing, variety, and drying conditions all may affect colour retention (se Figs 9.6 and 9.7). Of these, variety and storage temperature have the greatest influence on colour retention. The initial colour of pepper fruits at harvest or after dehydration is not a good indication of the rate of colour loss in storage (Lease and Lease, 1956). Therefore, varieties should be bred and evaluated for both initial colour and colour retention properties.

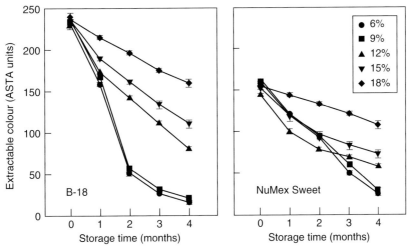

Fig. 9.6. Extractable colour of paprika samples with different pre-storage moisture contents and stored at ambient temperature (19–24°C) and humidity (24–42%) for 4 months. (Each point is the mean of four replications ± standard errors.) Source: Osuna-Garcia and Wall (1998).

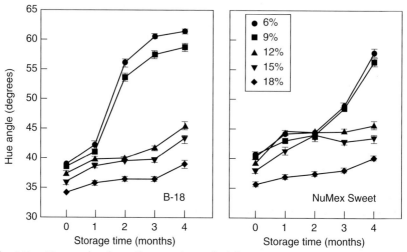

Fig. 9.7. Hue angle of paprika samples with different pre-storage moisture contents and stored at ambient temperature (19–24°C) and humidity (24–42%) for 4 months. (Each point is the mean of four replications ± standard errors.) Source: Osuna-Garcia and Wall (1998).

Carotenoid deterioration occurs during storage. Moisture content, storage temperature, and atmospheric composition are critical factors in the maintenance of pigment intensity (Chen and Gutmanis, 1968; Kanner *et al.*, 1977; Lee *et al.*, 1992). Adjusting the initial moisture content at which the dehydrated pepper product is stored could be an inexpensive method to delay colour loss. Osuna-Garcia and Wall (1998) demonstrated that colour loss can be minimized during storage at ambient temperature and humidity by increasing the percentage moisture content. For arid regions, a 15% pre-storage moisture content level may reduce colour loss of stored ground product by at least 50%.

A ripening agent, ethephon (2-chloroethyl phosphoric acid, Rhône-Poulenc Co.), can be used to hasten the ripening process. Defoliants or desiccants, such as sodium chlorate, are often used to both accelerate fruit drying during wet weather and aid in harvesting. Ethephon as a ripening enhancer may defoliate, as well as hasten maturity. This chemical will also increase the colour of red peppers that are harvested before frosts.

MEASURING COLOUR

Pepper colour can be evaluated from three different perspectives: surface colour, extractable colour and carotenoid profiles. Surface colour is a measurement of the visual colour perceived by the viewer. It is sometimes referred to as reflective colour. Surface colour varies according to cultivar, growing conditions, dehydration and storage conditions, and the coarseness of ground samples. Surface colour measurements are important when dehydrated pepper is to be used as a retail spice or as a coating on foods. Extractable colour is a measurement of total pigment content. Extractable colour analyses are useful when pepper is added as an ingredient or colourant in oil-based foods, cosmetics or pharmaceuticals. Extractable colour and surface colour measurements are standard quality evaluations in the spice industry. Analytical methods that separate and quantify individual pepper carotenoids, providing pigment profiles, are used mostly for research and development. HPLC is the most accurate method and is being used increasingly by oleoresin, drug and vitamin manufacturers for routine analysis.

Extractable colour is measured by a spectrophotometric process, and is designated ASTA units (American Spice Trade Association, 1985). Generally, the higher the ASTA colour value, the greater the effect on the brightness or richness of the final product. A pepper powder with 120 ASTA colour units would give a brighter red to a finished product than an equivalent amount of an 80 ASTA colour. Another term to describe red colour in oleoresin is the standard international colour unit (SICU), where 100,000 SICU is equal to 2500 ASTA units.

Surface colour measurement of pepper powder or fruits is based on the Hunter L.a.b. method using a colour difference meter (Conrad *et al.*, 1987). The '*L*' is the degree of whiteness–darkness on a scale of 100 to 0; '*a*' measures red when positive and green when negative; and '*b*' measures yellow when positive and blue when negative. Cultivar, stage of development at harvest, granulation and processing are all contributing factors to the final appearance of pepper. ASTA colour (extractable) affects the brightness of a product, while the surface colour has an impact on the hue of the product. Hue sets the kind of colour, e.g. brownish-red, orange-red or red-red. An orange-red cultivar can have a high level of red and yellow pigments, giving a high ASTA reading but a low L.a.b. reading.

Pepper fruit should be harvested when the pods have partially dried on the plant for maximum colour (Lease and Lease, 1956). Red, succulent pods have not fully developed their colour, whereas pods harvested late in the season have higher incidences of pod rots and sunburn. Storage temperature affects colour loss more than any other environmental factor (Lease and Lease, 1956). Colour loss is accelerated as temperature increases, and any exposure to light or oxygen hastens the rate of pigment bleaching. Storage at 3–5°C in the dark is recommended, although pepper powder removed from cold storage can lose colour quickly, reducing the retail shelf-life. Antioxidants such as ethoxyquin can be added to the product to reduce colour loss (Lease and Lease, 1956; Van Blaricom and Martin, 1951).

Ground pepper loses colour faster than whole pods during storage. For this reason, red pepper is often stored as flakes before final grinding. Flakes require less storage space than whole pods and maintain colour better than ground powder. The presence of seeds in ground pepper reduces the initial colour of the product but may actually decrease the rate of colour loss in storage. In one report, initial carotenoid content was highest in whole pods and coarse powder without seeds. However, during storage, coarse powder with seeds retained colour best (Lee *et al.*, 1992). Conversely, Lease and Lease (1956) reported that removing the seeds from red pepper powder had no effect on colour retention.

OLEORESIN

Oleoresin prepared from peppers is popular among food processors and other industries where a concentrated pungency or red colour additive is needed. When pungent peppers are used in the extracting process, the product is called 'oleoresin capsicum'. This product is used in medicinal and food industries. When non-pungent (paprika) pepper is used, the product is called 'oleoresin paprika'. Oleoresins are available in two basic forms: oil soluble or water soluble. Oleoresin is obtained from dried pepper pericarp by extraction with a volatile non-aqueous solvent (often hexane), which is subsequently removed from the oleoresin by evaporation at moderate temperatures and under partial vacuum. Oleoresins contain the aroma and flavour of the paprika or other pepper type, in concentrated form, and are usually viscous liquids, or semisolid materials.

Oleoresins are used for standardizing the pungency, colour and flavour of food products. Oleoresins are used because these traits are concentrated. Because of their high concentration, oleoresins cannot be incorporated into food products unless they are diluted. The dilution is usually achieved by dissolving the oleoresin in an appropriate solvent to make an essence. The paprika oleoresin is usually diluted with soybean oil. Oleoresin capsicum is made from the most pungent pod types and where colour content is not important. The oleoresin has a very high pungency and is used where a concentrated pungency is needed. This ranges from pharmaceutical uses to anti-mugger sprays, and has limited use in food products to modify the pungency level of a product.

Oleoresin extractors are located in many countries around the world. However, the facilities for producing high pungency oleoresin are limited to India, Africa and China. These facilities are close to the production areas for low cost, very pungent pepper pods. Paprika oleoresin is produced in several countries of which Spain, Ethiopia, Morocco, Israel, India, the USA, Mexico and South Africa are the top producers.

Pungency

There is a preference for specific levels of pungency in internationally traded pepper products and, with paprika, the absence of pungency is important.

The quality of red pepper and paprika products is based on pungency level, extractable red colour and flavour. Pepper must be processed and stored correctly to maintain quality. The dried red powder is classified into five groups based on pungency level: non-pungent or paprika (0–700 Scoville heat units), mildly pungent (700–3000), moderately pungent (3,000–25,000), highly pungent (25,000–70,000) and very highly pungent (> 80,000).

10

DISORDERS, DISEASES AND PESTS

Diseases and pests are primary constraints to pepper production. Peppers are susceptible to several diseases and pests which can reduce yield and quality of fruit. Not all the diseases and pests occur in the same region or at the same time. However, every region has specific diseases and pests which are of major importance in reducing pepper yields.

Disease and pest control is one of the most important factors in producing a profitable crop of peppers (DeWitt and Bosland, 1993). The correct diagnosis of a disorder is important in order to choose the proper treatment. A wrong diagnosis means selecting the wrong treatment which is expensive and unnecessary. A management strategy to control diseases that is compatible with the farming practices of the region makes the most sense. As a general rule, most pests cannot be eradicated, but they can be managed so that the risk of occurrence or loss is minimized. Pests are best controlled by taking action before they become serious. Frequent examination of pepper plants helps to diagnose potential problems. After a pest problem is well established, it is usually difficult to control.

Disease and insect control must start before pepper plants and seeds ever reach the field. A long-range programme is essential. Crop rotation is one of the best ways to promote healthy pepper production. Crop rotation helps minimize disease, especially root rot diseases caused by soil-borne pathogens. Proper plant spacing to provide adequate air movement around plants helps reduce foliar disease severity. An equally important disease control method is planting disease-resistant pepper cultivars. It is considered the most prudent means of disease control because of its low cost, ease of use and protection of the environment. In addition, planting healthy seeds and transplants, controlling water in the root zone, controlling insects that vector disease, and using sanitation techniques, such as cleaning and disinfecting equipment, will aid in producing an economical crop of high-quality peppers.

Pesticides can be under stringent restrictions of licensing, registration and use. Before a pesticide is applied to peppers, its package label should be reviewed

162

to determine if it is allowed. *Do not use **any** chemical if it is not labelled for peppers.* Furthermore, if the pepper is to be shipped to another country, be cognizant of any additional restrictions the receiving country may have. Many countries have a zero tolerance for certain pesticide residues.

Non-living (abiotic) and living (biotic) agents can cause pepper disease and injury (Black *et al.*, 1991). Non-living factors that cause disease include extreme levels of temperature, moisture, light, nutrients, pH, air pollutants and pesticides. Living pathogens include bacteria, fungi, mycoplasmas, virus, insects and nematodes. Descriptions and symptoms of abiotic disorders, bacterial diseases, fungal diseases, viral diseases and pests affecting peppers are listed below.

ABIOTIC DISORDERS

For peppers to grow optimally, they need a suitable temperature, good nutrition and the absence of stress factors. Abiotic disorders are a source of stress factors which occur in the absence of biological organisms. Some abiotic disorders can be caused by the lack of a major nutrient, e.g. nitrogen and potassium, or by the excess abundance of some elements, e.g. aluminium, boron or copper. Abiotic disorders may be similar in appearance to a biotic disease, and thus require a more careful investigation into the cause of the problem. Air pollution and salt injury are two abiotic disorders that are becoming more prevalent in pepper growing regions. Some abiotic disorders can be controlled by avoiding extreme temperatures, inferior soils and air pollutants. A variety of other improperly employed agricultural practices may cause damage to pepper plants. These include too deep cultivation, too much fertilizer or pesticide during an application, or an application of a chemical at the wrong time.

Air Pollution

Pepper is very susceptible to peroxyacetyl nitrate (PAN). PAN usually affects the underside of newly matured leaves (Maynard and Hochmuth, 1997). The PAN injury is confined to only three or four rapidly expanding leaves of sensitive plants. Very young leaves and mature leaves are highly resistant. The area becomes bronzed, glazed or silvery in appearance. Pale green to white areas may also appear on the leaf surfaces. The only control measure is not to grow in areas known to have dangerous levels of air pollutants.

Peppers are also reported to be highly sensitive to sulphur dioxide and have an intermediate sensitivity to hydrogen sulphide. The sulphur dioxide symptoms can be characterized as acute injury by dead tissue between the veins or on leaf margins. Chronic injury is characterized by leaves with brownish-red or bleached areas. Young leaves seldom show damage, while fully expanded leaves are very sensitive to sulphur dioxide.

Blossom-end Rot

Blossom-end rot occurs when the plant is unable to translocate adequate calcium to the pod, a condition caused by fluctuating soil moisture (drought or over-watering), high nitrogen fertilization or root pruning during cultivation. Wilting, lack of soil moisture and lack of calcium encourage the problem. This disorder first appears as a water-soaked area on the fruit. The tissue near the blossom end of the pods has a brown discoloration. Unlike tomatoes, blossom-end rot is never actually at the blossom end with peppers. Spots elongate and become brown to black, dry and leathery. Discoloured tissue shrinks until the affected area is flat or concave. Blemishes range from 0.5 cm spots to 8 cm long elongated spots. Pods affected with blossom-end rot usually ripen prematurely. Fungi growing on and within the infected pods are common. The fungi, however, are not the cause of the initial problem. Preventive measures include maintaining a uniform supply of soil moisture through irrigation and avoiding large amounts of nitrogen fertilizer. If manure is applied, turn it under in the autumn (as early as possible) so it will be well rotted before planting time. Irrigate when necessary during rapid pod development.

Oedema

This appears as numerous small bumps on the lower side of the leaves, and sometimes on the petioles. The cause is most likely to be over-watering. High humidity can also contribute to the cause. The control measures include reduced watering and better air circulation around the plant.

Flower and Bud Drop

Flower buds, flowers and immature pod drop is caused by a variety of conditions. Heat stress, insufficient water and excessive or deficient nutrient levels have been reported as causal agents. The best protection is to avoid over-fertilizing and under-watering. When the condition is corrected, the plant will resume flowering and fruiting. Turner and Wien (1994) reported that cultivars differ in their susceptibility to stress-induced flower abscission (drop). They found that susceptible cultivars reduce assimilate partitioning to flower buds and maintain high assimilate partitioning to expanded leaves.

Herbicide Injury

A hormone-type herbicide such as 2,4-D can cause distorted leaves. Other herbicides may cause chlorosis, necrosis or lesions. Control spray drift of herbicide application.

Mutations

Symptoms include leaf distortion, variegation in leaves and fruit deformity. Mutations can be mistaken for herbicide damage or viral infection. No control is possible and because very few plants manifest the mutation in the field, mutations are not of much economic impact. Seeds of interesting mutants can be sent to the Capsicum Genetics Cooperative, Box 30003, Dept 3Q, NMSU, Las Cruces, NM 88003, USA.

Salt Problems

High salt in the soil will 'pinch off' young seedlings at the soil line. A young seedling can die when light rains move the salt to the young, tender roots. Control salt problems by avoiding planting in fields with severe salt problems. Irrigate heavily prior to planting so that the salt is moved below root areas. Guttation-salt injury on peppers has been observed on the surface of pepper leaves following guttation through the adaxial surface, accompanied by injury in the form of large necrotic areas, often with a water-soaked border. It has been our observation that *Capsicum chinense* seems to be more susceptible to salt injury than *Capsicum annuum*.

Stip (Blackspot)

A physiological disorder causing grey-brown to greenish spots on fruit, most noticeable on red fruit that matures in the autumn (Smith *et al.*, 1996). The disorder only affects some pepper cultivars; this disorder manifests itself when peppers are grown under cooler temperature conditions and is associated with a suspected calcium imbalance and possible shorter day length. Stip can occur in the interior tissue of fruit as well as on the external surface. The best control is to plant resistant cultivars; 'Yolo Wonder L' and 'Grande Rio' have been identified as susceptible cultivars, while 'King Arthur' and 'Galaxy' are resistant cultivars.

Sunscald

Sunscald is caused by too much sunlight on fruit that has been growing in the shaded canopy. The smaller-podded varieties with erect fruits are not as susceptible to sunscald as are the large-podded varieties, such as bells and New Mexican pod types. Mature green fruits are more sensitive than mature red fruits. Symptoms are seen as a necrotic or whitish area on the fruit, on the side exposed to the sun, usually the afternoon sun. Often, fungi such as *Alternaria* spp. grow on affected areas of pods. Keep pods shaded by the plant's leaves or by screening. It is important during harvest to reduce damage to plants, i.e. removing too many leaves, because the remaining exposed fruit will sunscald. The best control is to avoid stress to the plant.

Wind Injury

In most cases, pepper plants can withstand moderate winds without significant injury. However, some larger plants may snap off at the soil line, where callus tissue has formed from wind whipping the plant back and forth in hard, crusty soil. Control the injury by erecting wind screens.

BIOTIC DISORDERS

Plant pathogens and pests are one of the most common causes of reduced yields in peppers. Pepper diseases have common names such as root rot, foliar blight, fruit rot, etc., but the most useful criterion for describing a disease is by the organism causing it.

Bacteria

There are about 1600 species of bacteria identified with many more thousands yet to be described. Most are saprophytic and do no harm to peppers. Bacteria are simple microorganisms usually consisting of single prokaryotic cells (Agrios, 1978). The cells contain a single circular chromosome, but no nuclear membranes or internal organelles, e.g. mitochondria or chloroplasts. In fact, bacteria and cellular organelles of eukaryotes are so similar that antibiotics that affect bacteria also inhibit mitochondria and chloroplasts, causing a leaf yellowing of the treated plants.

Bacterial spot (*Xanthomonas campestris* pv. *vesicatoria*)
Bacterial spot may be the most serious bacterial disease affecting peppers. There are currently seven races of the bacterium identified and no commercial

cultivar has resistance to all seven of the races (Sahin and Miller, 1996). However, the gene *Bs2* does provide resistance to races 0, 1, 2 and 3.

On young leaves, it causes small, yellowish-green to dark-brown coloured, raised spots. On older leaves the spots are dark, water-soaked and not noticeably raised. When spots are few, they may enlarge to 3 or 6 mm in diameter. The spots appear angular because the bacteria spreads along the veins. These spots have dead, straw-coloured centres with a dark margin. Severely spotted leaves turn yellow and drop. Infected seedlings often lose all but the top leaves of the plant. As disease progresses, the spots may enlarge, turn black, and become rough, giving the fruit a scabby appearance.

Management of bacterial spot depends on a combination of practices including the use of 'pathogen-free' seed and seedlings, sanitation, crop rotation, resistant cultivars and chemical applications. The best control is crop rotation and using disease-free seed. The organism is seed-borne and, in some areas, can overwinter on diseased plant refuse in the soil. Infected seedlings carry the disease to the field, where it can spread rapidly during warm, rainy weather, especially when driving rain and wind have caused injuries to the plants. Fixed copper compounds are commonly used to help manage the disease, although they are not highly effective under environmental conditions optimal for disease development or when high inoculum levels are present. Use of streptomycin on transplants is allowed in seedbeds, but not usually in greenhouses. The bacterium has developed resistance to both chemicals. This limits the efficacy of these compounds.

Before the current planting season, all seed should be treated by soaking in 1.31% sodium hypochlorite (one part liquid household bleach solution (5.25%) to four parts water) for 40 min with agitation (Goldberg, 1995). One litre of solution treats 0.5 kg of seed. Rinse thoroughly and dry seed promptly. Treated seed must be planted within the growing season because the treatment does affect seed longevity.

Bacterial canker (*Corynebacterium michiganense*)
The bacterium produces scabby canker spots on pepper pods (Volcani *et al.*, 1970). The spots coalesce to form large spots, 1–3 cm in diameter. It may also produce local lesions on fruits, stems and leaves, but does not induce systemic infection on the plant (Lai, 1976). In Israel, it is reported that the disease was mainly a problem in greenhouse production and in the field where crops are grown under cover. There is a possibility that infested pods can contaminate the seed with the bacterium.

Bacterial soft rot (*Erwinia carotovora* pv. *carotovora*)
Bacterial soft rot of pepper causes a soft rot of the fruit. The internal tissue softens and the pod turns into a watery mass. The pod also has a foul smell. The disease is most frequent when the weather is hot and humid. This postharvest rot can be devastating after harvest, in transit and in the market. Copper sprays

prior to harvest during hot, wet weather will reduce disease losses. Field infections are best controlled by keeping insect damage to a minimum. If the fruit is washed after harvest, the water should be chlorinated. Harvested fruit should never be washed in a tank of water unless the chlorine level in the wash water is maintained at 50 ppm. Keeping peppers cool, below 21°C, is also helpful. The disease can also be started by insect injury; therefore, controlling insects helps prevent this disorder.

Bacterial wilt (*Pseudomonas solanacearum*)

Bacterial wilt begins with a wilting of leaves. After a few days, a permanent wilt results, with no leaf yellowing. A test for this bacterium is to cut the roots and lower stems and look for an exudate of milky streams of bacteria when suspended in water. The best control is to plant clean seed and transplants.

Fungi

Fungi are one of the largest groups of organisms causing diseases on pepper. Peppers are susceptible to a number of fungi. The most important in terms of economic damage worldwide are discussed below. Several fungi that are important in a locale, but minor in the world are not discussed, e.g. choanephora blight and *Phyllosticta capsici*.

Anthracnose (*Colletotrichum* spp.)

Anthracnose is the common name for the disease caused by several species of the fungus *Colletotrichum*. Symptoms are small, water-soaked, shrunken lesions that expand rapidly. The disease is most serious on ripe pods. The lesions have dark fungal spores in them, and a characteristic concentric ring (target-shaped spots) appearance. Clean seed and crop rotation are important. Fungicides may be helpful for control.

Early blight (*Alternaria solani*)

Early blight causes damage to the leaves and the fruit. The disease appears as small, irregular, brown, dead spots usually found on older leaves. The spots enlarge until they are 6–12 mm in diameter. The spots are ridged and have a target pattern. Early blight is usually more prevalent than anthracnose.

Cercospora leaf spot (*Cercospora capsici*)

This disease is also called 'frogeye'. A leaf or stem lesion is oblong or circular with a small, light-grey centre and dark-brown margin (frogeye). The diseased spots usually dry and fall from the leaf, leaving conspicuous holes. Leaf drop is common with severely infected leaves. Stems and fruits are especially susceptible to this disease which is worst under humid conditions. The fungus is active during the same environmental conditions that favour bacterial spot. In fact, these two diseases, along with *Alternaria* are often found together on infected leaves. Because the disease is seed-borne, planting clean seed and crop rotation are important in controlling the disease. Fungicides can help to manage the disease.

Damping-off/seedling disease

Several fungi, such as *Pythium, Rhizoctonia, Fusarium*, etc., are associated with this disorder. Seedlings fail to emerge (pre-emergence damping-off), small seedlings collapse (post-emergence damping-off) or seedlings are stunted (root rot and collar rot). In the field, seedling diseases develop during cold, wet periods. *Pythium aphanidermatum* infections can produce symptoms similar to phytophthora root rot and pod rot (Apple and Crossan, 1954). Damping-off is also a problem in raising transplants in the greenhouse, where high humidity and frequent overhead watering can favour seedling diseases. Additional causes of seedling loss include poor seed quality, improper planting depth, high salt concentrations, wet seed bed, strong winds, severe nutrient deficiencies or toxicity, pre- and post-plant herbicide applications and insects.

In order to prevent seedling diseases, plant only high quality seed or transplants, avoid fields and seed beds that are poorly drained. A vigorously growing seedling is the best protection against damping-off/seedling disease. Treating the seed with a fungicide, or treatment of soil to minimize the fungi population in the soil will protect against pre-emergence damping-off. Good air circulation is important, so fans in greenhouse situations are helpful.

Fusarium stem rot (*Fusarium solani*)

Fusarium stem and fruit rot has been reported on greenhouse peppers. *Fusarium solani* is the causal agent. The disease was mostly found infecting the nodes of the pepper plant (Fletcher, 1994). The fungus causes soft dark lesions. Pepper fruit may also become infected around the calyx especially if it is damaged or ripe. The tissue becomes dark and sunken, with tiny red spheres or orange/pink spore pustules forming at these infected sites.

Grey mould (*Botrytis cinerea*)

The fungus causes a sudden collapse of succulent tissues, such as young leaves, stems and flowers. Grey powdery spore masses of the fungus occur on the surface of dead plant tissues. High humidity favours the disease. A wide plant spacing so that plants dry quickly helps reduce the disease. A fungicide may be used if the mould is severe.

Phytophthora (*Phytophthora capsici*)

This water mould can invade all plant parts. The fungus may cause at least three separate disorders: foliar blight, fruit rot and root rot (Bosland and Lindsey, 1991; Alcantara and Bosland, 1994). It spreads rapidly when humidity and temperatures are high and/or the soil is wet. The first symptom of phytophthora root rot is severe wilting. Within days the plant is dead. Phyto-phthora root rot can be prevented by avoiding excess water in the field. One cultural control measure is cultivating so that plants are grown on a high ridge to allow water to drain away from the roots; another is to irrigate alternate rows. Some chemical fungicides have been shown to be effective against foliar blight and pod rot, but not against root rot.

Powdery mildew (*Leveillula taurica-Oidiopsis taurica*)

The disease is favoured by warm temperatures (20–35°C). Although high humidity favours germination of spores, infection can occur during periods of high or low humidity. Wind-disseminated spores cause secondary infections. On the adaxial leaf surface, chlorotic blotches or spots, which may become necrotic, are symptomatic. On the abaxial leaf surface, a white to grey powdery growth may exist. Infected leaves may drop prematurely. Fungicides have been effective in controlling the disease. Resistant cultivars when available will also be useful.

Rhizoctonia root rot (*Rhizoctonia solani*)

Rhizoctonia solani infects a large number of hosts and can cause disease in seedlings and mature plants. *Rhizoctonia* infection is thought to occur in the spring at seedling age. Rhizoctonia is not an aggressive pathogen. A cool and damp environment is optimal for infection. The symptoms may occur when the plant is under heat or water stress, and include wilting and death of the plants scattered throughout the field. The taproot will have reddish-brown lesions which are a diagnostic characteristic for this disease. Seed treatment with a fungicide and crop rotation are the best control measures. There are no resis-tant cultivars, but some accessions do have resistance (Muhyi and Bosland, 1995).

Stemphylium leaf spot (*Stemphylium botryosum* f. sp. *capsicum*)

The organism causes leaf spots up to 3–4 mm in diameter, eventually resulting in a severe defoliation of the lower and middle leaves of the diseased plant (Braverman, 1968). The fungus is a ubiquitous species and may become established in many kinds of environments. There appears to be specialization in the fungus, with the pepper isolate infecting pepper and tomato, but not lucerne.

Southern blight (*Sclerotium rolfsii*)

Southern blight usually causes disease during the hot and wet season. It causes plants to wilt as a result of stem girdling and rot at the soil surface. Occurrence is usually in islands or 'hot spots' in the field, with a few infected plants scattered throughout the field. The base of the stem is brown and decayed above and below the soil line. White fungus is visible at the base of the stem and on the soil around the base. Sclerotia, small brown spheres about the size of mustard seeds, can be found in the fungus. Deep tilling to bury the sclerotia, removing infected plants and allowing a plot to set fallow for a couple of years are helpful control measures. Soil fungicides may provide some control.

The control procedure is a combination of tactics: (i) rotating with maize or small grain; (ii) ploughing deeply to bury crop residues; (iii) using a fungicide in transplant water; (iv) shallow cultivation which does not throw soil up on the plants; (v) removing infected plants promptly; (vi) fumigation of the field; and (vii) avoiding problem fields.

Verticillium wilt (*Verticillium dahliae*)

This disease is caused by a soil-borne fungus that is primarily a problem in temperate climates. Symptoms of verticillium wilt are highly variable. Plants may show a yellowing of leaves and stunting. As the disease progresses the plants can shed leaves and may finally die. If the stem is cut, a vascular discoloration is seen. No resistant cultivars nor chemical controls are known. The most effective control is to avoid fields where verticillium wilt has been observed. Crop rotation with small grains may help to reduce the pathogen population in the soil.

White mould (*Sclerotinia sclerotiorum*)

White mould, or sclerotinia disease, causes a wilt, rot and blight. Blighting or rotting of any above-ground or below-ground plant parts can occur. At first, the affected area of the plant has a dark-green, greasy or water-soaked appearance. On stems, the lesion may be brown to grey in colour. If the humidity is high, a white, fluffy mycelial (mould) growth appears. Lumpy areas appear in this white growth, which become hard and black as they mature. The hard, black bodies (sclerotia) form inside the stem or on the outside surfaces of the stem and other plant parts. Control includes well-drained soil, proper plant spacing, crop rotation and careful removal of all infected plants as soon as possible.

Mycoplasma

Stolbur

Stolbur is caused by a mycoplasma. Mycoplasma, along with bacteria and rickettsiae, are prokaryotic organisms. Most mycoplasma are carried from plant to plant by insects. The early symptoms of stolbur are a scalded look to the plants with limp and yellow leaves. A diseased plant seldom sets fruits. Leaves drop off from the top of the plant downward. The vector is a cicada which, once infected, will stay infected throughout its life. There are no resistant cultivars. The best control is to keep weeds that may harbour the mycoplasma under control and control the cicada population.

Viruses

In the tropics, viruses are the most serious disease problem of peppers. Some 45 viruses have been reported to infect peppers (Green and Kim, 1991). Of these, more than half are transmitted by aphids. The other viruses are transmitted by nematodes, thrips, leafhoppers, whiteflies, beetles, fungi or by the grower handling infected plants. Several are transmitted by mechanisms not yet understood. Viruses alter the metabolism of plant cells, causing the plants to grow abnormally. This condition causes both decreased yields and visible symptoms, such as distorted leaves, abnormally coloured leaves, dead tissue, mottled or mosaic leaves or fruit, stunted plants or curled leaves. One plant can be attacked by many viruses and may express many different symptoms.

Avoid using tobacco in any form to help protect plants against tobacco mosaic virus. Growers and farm labourers who use tobacco should wash their hands with soap and water or rubbing alcohol before handling healthy plants. Early detection and removal of infected plants helps, but complete control is often difficult. Symptoms of virus infection vary greatly in expression and severity, and include mild mottle, mosaic, veinbanding, ringspots, various types of necroses, leaf discoloration, deformation and blistering, and severe stunting of the whole plant. Leaves stems, flowers and fruits may all be affected.

Mosaic virus symptoms have intermixing of light- and dark-green areas on the leaves. The mottled areas have irregular outlines, and may follow the main veins. Infected leaves are generally smaller than healthy leaves, and are often slightly puckered and have curled edges. In severe cases, the leaves may become long, narrow and twisted. Infected plants are usually more dwarfed and bushy than healthy plants and have reduced yields. Separating symptoms caused by mosaic diseases from those caused by abnormal pH, herbicide injury, nutritional deficiencies, feeding damage by mites or insects and so forth, may be difficult. Viruses are hard to control. No viricides exist that control plant viruses. To help reduce mosaic virus, use these cultural methods: plant virus-free seed, remove weeds, control insects, remove plants showing virus symptoms and plant resistant varieties. Most pepper viruses are distributed worldwide with the exception of chili veinal mottle virus, pepper severe mosaic virus, pepper veinal mottle virus, pepper mild mosaic virus and pepper mottle virus. These have been reported only in certain geographic areas.

For devising effective control measures and to initiate efficient breeding programmes it is important that the viruses present in the particular geographic area are correctly identified and characterized and that their epidemiological behaviour is understood. For the prevention or reduction of virus infection, particularly for the viruses transmitted by aphids in a non-persistent manner, the following practices have been tried with varying degrees of success in grower's fields: organic mulches; aluminium foil strips above the crop; insect traps; mulches of aluminium foil, silver vinyl, or white or translucent polyethylene; aluminium painted polyethylene sheets; sticky yellow polyethylene sheets; mineral oil sprays; skimmed milk sprays; white washes and the cultivation of non-susceptible barrier crops such as maize.

Spraying pepper fields with 1.25% and 2.85% concentrations of a light oil emulsion at intervals of 5–6 days reduced the field spread of viruses, mainly potato virus Y (PVY) (Nitzany, 1966). However, the delay in virus spread did not result in higher yields. Further studies indicated that oil spraying was highly effective for control of aphid-borne cucumber mosaic virus (CMV) and potato virus Y (Loebenstein *et al.*, 1970). The oil sprays were less effective against late infection of mature plants. The virus infections were delayed, but increases in yield were small. The research suggests that spraying to protect young seedlings from virus infection is warranted, but field spraying is of questionable use.

The use of 'coloured baits' has been effective in controlling the spread of aphid-transmitted viruses (Loebenstein *et al.*, 1970). The spread was reduced by using sticky sheets of yellow polyethylene located outside the field (Cohen and Marco, 1973). This trapped the winged form of the aphids. This method reduced the spread of CMV and PVY in the field. Antiviral agents such as cytovirin have also been tested for the control of pepper viruses but were found to be phytotoxic to the plants themselves at doses that were most effective. (Simons, 1960). The planting of resistant cultivars is the best way to control viruses. Many virus-resistant cultivars have been released.

Alfalfa mosaic virus (AMV)
AMV is aphid-transmitted in a non-persistent manner and produces a distinctive white calico pattern on the leaves. Peppers planted near lucerne fields have a higher incidence of the disease. For control, reduce aphid populations and avoid planting near lucerne fields.

Beet curly top virus (BCTV)
Curly top virus (CTV), or beet curly top virus (BCTV) as it is more formally known, is widespread throughout arid and semi-arid regions of the world. The virus is common in the western United States from Mexico to Canada and in the eastern Mediterranean Basin. More than 300 species in 44 plant families are susceptible to this virus. The virus is a geminivirus, a ssDNA virus. The most striking symptom is stunted and yellowed plants. The plants are also quite stiff and erect, and the leaves have a leathery feel. The virus vector is beet leafhoppers (*Circulifer tenellus*). The leafhopper is an effective vector because it is able to transmit the virus after feeding on an infected pepper plant for as little as 1 min and can subsequently transmit the virus for the remainder of its lifetime. Fortunately, the virus is not passed on to leafhopper progeny. Leafhoppers that carry the virus do not feed in shady locations. Spraying or dusting with insecticide can be justified only when control is needed for other insects. If possible, all diseased pepper plants should be removed from the field as soon as they are noticed so that they do not continue to provide a source

of virus for transmission to heathy plants. Resistant germplasm has been identified and may be useful in breeding cultivars resistant to the disease (Ungs *et al.*, 1977).

Cucumber mosaic virus (CMV)

CMV is one of the most important virus diseases of peppers worldwide. The virus exists as a number of strains, but all are apparently capable of infecting pepper and differ only in symptom expression. The age of the pepper plant at the time of infection strongly influences the types of symptoms to be manifested. CMV symptoms appear often on lower, mature leaves as ring-spot or oak-leaf necrotic patterns. Ring-spot symptoms are more prominent on determinate-type peppers like bell peppers. The necrotic symptoms, whether they occur on the foliage or on the fruit, are basically a shock reaction attributed to early virus infection. Sometimes adjacent pepper plants display only a mild to moderate mosaic pattern and have a general dull appearance. This difference may be influenced by the particular CMV strain involved, but more likely reflects the age at which plants are infected. With early infection, both quality and quantity of fruit production will be affected.

CMV is carried by aphid species in a non-persistent manner. Strategies to delay early infection should be used to enhance yield and reduce the number of culled fruit. Isolate pepper plantings from weedy border areas or grow them next to taller border plantings, such as maize, which can function as a non-susceptible barrier crop. Mineral oil sprays have been used to interfere with the transmission of all pepper viruses by aphids.

Geminiviruses

In Mexico and Texas, several 'new' viruses have been detected. The local name for the viruses are chino del tomate (tomato crinkle), serrano golden mosaic, sinaloa tomato leaf curl, pepper mild tigre and Texas pepper virus. The viruses have similar symptoms, but are biologically distinct. The common symptoms are stunting, curling or twisting of the leaves; bright yellow mosaic; distorting of leaves and fruit; and reduced yield. The viruses are spread by whiteflies (*Bemisia tabaci*). Control of geminiviruses is difficult once plants become infected. Destruction of perennial weeds which harbour the whiteflies and crop rotation are currently the only control measures.

Pepper mottle virus (PeMV)

Plants infected with PeMV share many traits with PVY and tobacco etch virus, including veinbanding, but the mottling is more extensive in interveinal areas and over the entire leaf surface. Fruit mosaic and distorted fruit are also common symptoms. Aphids trasmit PeMV, and control practices include good sanitation and planting of resistant cultivars if available.

Potato virus Y (PVY)

PVY is a common virus among solanaceous crops, infecting potato and tomato in addition to pepper. The symptom most useful for diagnosing PVY infection is a mosaic pattern that develops along the veins, commonly referred to as veinbanding. Other symptoms include leaf distortion and plant stunting with early infection. Like CMV, PVY is transmitted by several aphid species, but the green peach aphid is generally considered to be the most important vector. PVY has a limited host range, so controlling solanaceous weeds bordering the crop reduces one potential source of inoculum. Resistant cultivars are becoming more commonplace.

Tobacco etch virus (TEV)

TEV normally occurs with the PVY virus. Symptoms include broad, dark-green mosaic bands along the veins, beginning at the leaf base and often continuing to the tip. Cultivars with PVY resistance help to control TEV because resistance is closely linked to PVY. An interesting interaction between TEV and tabasco plants is that instead of the mosaic pattern, tabasco plants wilt and die as if infected by a bacterium. Whole fields of tabasco can be lost. TEV is spread by aphids, but planting of resistant varieties is the best control.

Samsun latent tobacco mosaic virus (SLTMV)

Typical symptoms include mild mosaic and leaf distortion. Pods develop rings, line patterns, necrotic spots and distortion. Plant stunting may also occur. SLTMV is spread mechanically, by hands touching an infected plant and then touching an uninfected plant, so disinfecting hands helps. Clean seed and crop rotation also help to control the virus.

Tobacco mosaic virus (TMV)

TMV is generally not a problem for bell pepper production because most cultivars are resistant to the common strains of the virus. Mosaic and systemic chlorosis and leaf drop will occur on susceptible plants. TMV is spread mechanically, by hands touching an infected plant and then an uninfected plant, so disinfecting the hands with alcohol helps. Clean seed and crop rotation are the best prevention.

Tomato spotted wilt virus (TSWV)

TSWV is common in both temperate and subtropical areas of the world. Thrips transmit the virus, but only larvae can acquire the virus. Only adult thrips that fed on infected plants as larvae can transmit the virus and then only after a latent period of 4–10 days. This type of transmission is very different from aphid transmission. The virus causes sudden yellowing and browning of the young leaves which later become necrotic. Fruit formed after infection develop large necrotic blotches.

PESTS

The insects most common to pepper plants are cutworms, aphids, pepper weevils, maggots, flea beetles, hornworms and leafminers. Early in the season, cutworms are the most damaging pests to both seeded and transplanted peppers. Seeded peppers are also subject to attack by flea beetles when the cotyledons emerge. Green peach aphids can become numerous at any time, but are probably more prevalent during the summer. Besides the stress created by aphids feeding on plant sap, their honeydew gets on the fruit and leaves. Its presence on the leaves, if heavy enough, can decrease photosynthesis by sooty mould growth. Occasionally, loopers will feed on the foliage, exposing the pods to sunscald. Fall and beet armyworms, yellow-striped armyworms and variegated cutworms may feed on pods. The beet armyworm will also feed on the foliage. The corn earworm feeds on pods and causes the pods to drop or become unmarketable.

Problem insects differ in each region. To control the insect population and keep seedlings insect-free, inspect the plants daily, weed well around the peppers, dispose of diseased plants immediately and use insecticides if necessary. Remember that only approved pesticides should ever be used. Following are descriptions of the insects and other pests which most commonly attack peppers.

Insect Pests

Cutworms
Several species of cutworms exist; they are the larvae of a large family of moths. They are dull grey, brown or black, and may be striped or spotted. They are stout, soft-bodied, smooth and up to 5 cm long. When disturbed, they curl up tightly. Cutworms attack only seedlings. They cut off the stems above, at or just below the soil surface. Cultivation disturbs the overwintering places of the cutworm. A non-chemical control is to place cardboard, plastic or metal collars around the young plant stems, and push the collar 2.5 cm into the ground to stop the cutworms.

Green peach aphids (*Myzus persicae*)
These aphids are usually light-green and soft-bodied. They cluster on leaf undersides or on stems. Aphids excrete a sticky liquid called honeydew, which creates spots on the foliage. A black fungus, sooty mould, may then grown on the honeydew. Severe infestations can cause wilting, stunting, curling and leaf distortion. Usually, aphid predators and parasites keep the aphid numbers low, but the aphids can multiply quickly (Votava and Bosland, 1996).

European corn borers (*Ostrinia nubilalis*)

Corn borer moths are a key pest because they can be found in almost every field every year. To control the borers, the larval stage must be targeted. The eggs are deposited on peppers and, as they hatch, the larvae tunnel into the pepper pod. Eggs hatch 4–5 days after being laid and this is the most appropriate time for control measures. Pheromone traps and visual sighting of adult moth populations are two of the best ways to monitor the pest. It is recommended that the first spray begin within 7–10 days of initial moth emergence or within 4 days of heavy emergence.

Flea beetles (*Epitrix* spp.)

These black beetles are about 3 mm long. Young plants are severely damaged and full of holes. The flea beetle is repelled by shade.

Fruitworms (*Heliothis zea*, corn earworm; *Spodoptera* spp., armyworm, etc.)

Fruitworms include the fall armyworm, beet armyworm and tomato fruitworm (corn earworm). At the larval stage, the worms are green, brown or pink, with light strips along the sides and on the back. It grows to 4.5 cm long. The fruitworm damages the pods by eating holes in the fruits.

Grasshoppers

There are many species of grasshoppers that will eat pepper foliage. Adults have front wings that are larger than the body and are held roof-like over the insect. The hind legs are long and adapted to jumping. They may destroy complete plantings. Soil cultivation is useful because grasshoppers lay their eggs in the top 7 cm of soil.

Grubs (*Phyllophaga* spp.)

There are more than 100 species of grubs. They are white to light-yellow with dark-brown heads. They are curved and 1–4 cm long. White grubs are the larvae of may beetles. The grubs live in the soil and may take 3 years to mature. The grubs feed on roots and underground parts. Cultivation is a good control measure.

Hornworms (*Manduca sexta* and *Manduca quinquemaculata*)

The worms are the larval stage of the sphinx moth; these large caterpillars have a green body with diagonal lines on the sides and a prominent horn on the rear end. They can be up to 10 cm long. They ravenously eat foliage and can strip a pepper plant, killing it.

Leafhoppers

Leafhoppers are an important group of small, sap-sucking insects. There are many species of leafhoppers, but the leafhopper, *Circulifer tenellus*, spreads curly top virus. They are usually green, wedge-shaped and up to 3 mm long. They fly quickly when disturbed. Nymphs resemble the adults but are smaller. The leafhoppers can cause hopperburn but this is rare in pepper. The symptoms are tips and sides of pepper leaves turning yellow to brown and becoming brittle. Remove infested plants or plant parts immediately.

Leaf miners
Many species of flies will cause the leaf mining disorder. The larva is yellow, about 3 mm long and lives inside the leaves. The adult is a tiny (less than 3 mm in length), black and yellow fly. The infected leaves are blotchy. The larvae make long, slender, winding mines under the epidermis of the leaves. Remove infested leaves.

Pepper maggots (*Zonosemata electa*)
The maggot is the larva of a tephritid fly. The slender white or yellowish-white maggot is about 7–14 mm long. Adult flies are yellow-striped flies about 5 mm long with dark bars on the wings. Maggots feed within the pepper pod, causing it to decay or drop from the plant.

Pepper weevils (*Anthonomus eugenii*)
The pepper weevil is a severe pest in tropical areas and can cause damage in temperate regions when introduced. Adult pepper weevils feed on leaves, blossom buds and pods. The adult will lay eggs on the flowers, buds and fruit. The eggs hatch and the larvae burrow into the young pods, feeding inside the fruit. Premature fruit drop results. The larvae are white with brown heads. Whole fields have been abandoned because of fruit loss. Once the larvae are inside the fruit practical control is impossible. Control of pepper weevils is based on frequent and accurate scouting. The recommended threshold for pepper weevil is one adult per 200 plants. When the threshold is reached regular applications of insecticides are recommended. Cultural control includes destruction of crop residue and weeds of the nightshade group to reduce the possibility of adult weevils overwintering. Resistant cultivars are not known at this time.

Thrips
There are many species of thrips and all are extremely small (less than 1 mm long). The mouth parts are classified as rasping-sucking. They can produce a new generation every 2 weeks. Leaves are distorted and curl upward (boat-shaped). The lower surface of the leaves develops a silvery sheen that later turns bronze. Malathion, sulphur and diatomaceous earth are effective. The species, *Frankliniella tritici*, is the vector for tomato spotted wilt virus.

Stink bugs
There are many species of stink bugs, most having green, blocky bodies about 14 mm long. Both the nymphs and adults damage plants by sucking sap

primarily from the pods. Fruits have a cloudy spot, a whitish area with indistinct borders. Damage is similar to hail damage on fruit, but hail damage tears leaves.

Tarnished plant bugs (*Lygus lineolaris*)
These insects have a greenish to brown body 7 mm long. There are yellow, brown and black markings on the body. There is a yellow tinge at the end of each fore wing. They inject a toxin when feeding on blossoms and buds, causing them to drop.

Whiteflies
Whiteflies are minute insects (2 mm) with broad wings that are covered with a fine, white, waxy powder. The immature and adult stages suck plant juices from the leaves causing the leaves to shrivel, turn yellow and drop. In addition, they can transmit viruses to pepper plants. Whitefly control is difficult. Insecticides combined with good cultural practices, such as removing infected plants, is the best approach for control.

Other Pests

Spider mites (*Tetranychus* spp.)
Spider mites are red arachnids less than 1 mm in length. When infestation is high, the leaves will have webs on them. Leaves curl downwards (inverted spoon). A bronzed or russeted appearance on leaves or fruits occurs. They can kill the plant if left uncontrolled. It is best to treat spider mites when they are in the early stages. Miticides are marketed for their control. Repeated application is necessary and spider mites have been known to develop resistance to miticides, therefore alternative chemicals may have to be sprayed periodically.

Nematodes
Peppers are subject to attack by various nematodes. Among these are species of *Meloidogyne* and the root-lesion nematode, *Pratylenchus penetrans*. Nematodes are microscopic roundworms that feed on plant roots. Nematodes are not able to move from one field to another easily because of their tiny size and aquatic nature, and they do not usually survive in blowing dust. Nematodes are spread mainly by agriculture and related activities that inadvertently move soil around on equipment, transplants, the feet of humans and other animals, or in irrigation water. Three species of root-knot nematodes cause serious damage to peppers: *Meloidogyne incognita*, *Meloidogyne hapla*, and *Meloidogyne arenaria*. Symptoms of nematode injury to pepper plants vary with plant age and the severity of the nematode infestation. Because nematodes are sensitive to soil type, damage usually appears 'patchy' rather than uniform in a field. Above-ground symptoms include plant stunting and leaf wilting. Roots infected with

root-knot nematodes may have obvious swellings or galls. The galls vary in size from smaller than a pin-head to larger than a pea. Injury is more severe in sandy soils. The damage caused by nematodes can also lead to secondary infections by soil fungi. The older literature that recommends rotating with small grains to reduce damage is not reliable because host susceptibility was based on galling, not egg production. It is now known that some grains can be excellent hosts for root-knot nematodes without forming noticeable galls.

The symptoms of *Pratylenchus penetrans* infection are severe stunting, foliar chlorosis, and failure of or reduced fruit setting. Leaves are smaller and fewer. Roots are retarded in growth (without galls) and have brown to dark-brown lesions. The nematode causes mechanical destruction of parenchyma cells in the root cortex, but does not enter the stele. The use of nematicides has been the primary control. As with many other crops, the number of materials registered for use on peppers has declined significantly. Effectiveness of nematicides depends on soil texture, level of organic matter, soil temperature and soil moisture. Use of resistant cultivars will play an ever-increasing role in pepper production.

Animals

It is common in rural areas for herbivores such as deer, rabbits and mice to destroy peppers. In Ethiopia, gazelles are notorious for eating pepper plants. In the urban setting, numerous complaints come when dogs eat pepper plants in the home garden. At New Mexico State University, it has been observed that skunks will eat the fruits right from the plants. Birds can be pests by pecking holes in pepper pods to get at the seeds. The only solution besides killing the offender is to use fences and screenings.

CONCLUSION

The type and amount of losses in pepper caused by plant diseases and pests will vary with locality. Not only can pepper plants be killed by diseases and pests, but the quality of the peppers can be affected to the point that it is not economical to harvest even though the pepper plant has produced a substantial yield. Some pepper diseases and pests can be controlled by a specific measure, for example an application of fungicide. There are other diseases and pests where no measure exists as yet, for example curly top. However, with the continued introduction of new disease resistant cultivars and appropriate cultural measures, the production of abundant high quality peppers can be assured.

REFERENCES

Afek, U., Rinaldelli, E., Menge, J.A., Johnson, E.L.V. and Pond, E. (1990) Mycorrhizal species, root age, and position of mycorrhizal inoculum influence colonization of cotton, onion, and pepper seedlings. *Journal of the American Society for Horticultural Science* 115, 938–942.

Agrios, G.N. (1978) *Plant Pathology*. Academic Press, New York.

Albregts, E.E. and Howard, C.M. (1973) Effect of fertilization and mulching with bio-degradeable polyethylene-coated paper on responses of okra and pepper. *HortScience* 8, 36–38.

Alcantara, T.P. and Bosland, P.W. (1994) An inexpensive disease screening technique for foliar blight of chile pepper seedlings. *HortScience* 29(10), 1182–1183.

Alcantara, T.P., Bosland, P.W. and Smith, D.W. (1996) Ethyl methanesulfonate-induced seed mutagenesis of *Capsicum annuum*. *Journal of Heredity* 87, 239–241.

Alcorn, J.B. (1984) *Huastec Mayan Ethnobotany*. University of Texas Press, Austin, Texas.

Allard, R.W. (1960) *Principles of Plant Breeding*, 1st edn. John Wiley & Sons, New York.

American Spice Trade Association (ASTA) (1985) *Official Analytical Methods of the American Spice Trade Association*. Englewood Cliffs, New Jersey.

Andrews, J. (1984) *Peppers: the Domesticated Capsicums*. University of Texas Press, Austin, Texas.

Anghiera, P.M. (1493) Decadas de nuevo mundo. Cited in Andrews, 1984.

Apple, J.C. and Crossan, D.F. (1954) Pythium stem rot and wilt of pepper. *Plant Disease Reporter* 38, 555–556.

Armitage, A.M. (1989) Promotion of fruit ripening of ornamental peppers by ethephon. *HortScience* 24, 962–964.

Armitage, A. and Hamilton, B. (1987) Ornamental peppers: a hot new crop. *Greenhouse Grower* 5, 92–95.

Bailey, L.H. (1923) Capsicum. *Gentes Herbarum* 1, 128–129.

Batal, K.M. and Granberry, D.M. (1982) Affects of growth regulators on ripening and abscission of pimiento and paprika peppers. *HortScience* 17, 944–946.

Batal, K.M. and Smittle, D.A. (1981) Response of bell pepper to irrigation, nitrogen, and plant population. *Journal of the American Society for Horticultural Science* 106, 259–262.

Beckstrom-Sternberg, S.M., Duke, J.A. and Wain, K.K. (1994) The ethnobotany database. HTTP://probe.nalusda.gov:8300/cgi-bin/browse/ethnobotdb

Beese, F., Horton, R. and Wierenga, P.J. (1982) Growth and yield response of chile pepper to trickle irrigation. *Agronomy Journal* 74, 556–561.

Belletti, P.M.C., Nada, E. and Lanteri, S. (1995) *Flow cytometric estimation of nuclear DNA content in different species of Capsicum*. Eucarpia, Budapest, Hungary, pp. 22–25.

Bennett, D.J. and Kirby, G.W. (1968) Constitution and biosynthesis of capsaicin. *Journal of the Chemistry Society (C)* 442–446.

Betanzos, Juan de (d.1576) *Suma y Narración de los Incas*; translated and edited by Hamilton, R. and Buchanan, D. from the Palma de Mallorca manuscript as *Narrative of the Incas by Juan de Betazos* (1996). University of Texas Press, Austin, Texas. p. 171.

Bhargava, Y.R. and Umalkar, G.V. (1989) Productive mutations induced in *Capsicum annuum* by physical and chemical mutagens. *Acta Horticulturae* 253, 233–237.

Bhatt, R.M. and Srinivasa Rao, N.K. (1993) Response of bell pepper (*Capsicum annuum* L.) photosynthesis, growth, and flower and fruit setting to night temperature. *Photosynthetica* 28, 127–132.

Biacs, P.A., Czinkotai, B. and Hoschke, A. (1992) Factors affecting stability of colored substances in paprika powders. *Journal of Agriculture, Food and Chemistry* 40, 363–367.

Black, L.L. and Rolston, L.H. (1972) Aphids repelled and virus diseases reduced in peppers planted on aluminum foil mulch. *Phytopathology* 62, 747.

Black, L.L., Green, S.K., Hartman, G.L. and Poulos, J.M. (1991) *Pepper Diseases: a Field Guide. Asian Vegetable Research and Development Center*, Taipei, AVRDC Publication No. 91-347.

Bosland, P.W. (1992) Chiles: a diverse crop. *HortTechnology* 2(1), 6–10.

Bosland, P.W. (1993) An effective plant field-cage to increase the production of genetically pure chile (*Capsicum* spp.) seed. *HortScience* 28(10), 1053.

Bosland, P.W. (1997) *Capsicum: a Comprehensive Bibliography*, 5th edn. The Chile Pepper Institute, Las Cruces, New Mexico.

Bosland, P.W. and Gonzalez, M.M. (1994) 'NuMex Mirasol' chile. *HortScience* 29(9), 1091.

Bosland, P.W. and Iglesias, J. (1992) 'NuMex Bailey Piquin' Chile Pepper. *HortScience* 27, 941–942.

Bosland, P.W. and Lindsey, D.L. (1991) A seedling screen for Phytophthora root rot of pepper, *Capsicum annuum*. *Plant Disease* 75(10), 1048.

Bosland, P.W., Iglesias, J. and Bailey, A. (1988) *Capsicum Pepper Varieties and Classification*. Cooperative Extension Service-Circular 530. New Mexico State University, Las Cruces.

Bosland, P.W., Iglesias, J. and Gonzalez, M.M. (1993) 'NuMex Joe E. Parker' chile. *HortScience* 28, 347–348.

Bosland, P.W., Iglesias, J. and Gonzalez, M. (1994) `NuMex Centennial' and `NuMex Twilight' ornamental chiles. *HortScience* 29, 1090.

Boyette, M.D., Wilson, L.G. and Estes, E.A. (1989) *Proper Postharvest Cooling and Handling Methods. North Carolina Agricultural Extension Service Publication*, Raleigh, AG-414-1.

Boyette, M.D., Wilson, L.G. and Estes, E.A. (1990) *Postharvest Cooling and Handling of Peppers. North Carolina Agricultural Extension Service Publication*, Raleigh, AG-413-3.

Bradford, K.J., Steiner, J.J. and Trawatha, S.E. (1990) Seed priming influence on germination and emergence of pepper seed lots. *Crop Science* 30, 718–721.

Braverman, S.W. (1968) A new leaf spot of pepper incited by *Stemphylium botryosum* f. sp. *capsicum*. *Phytopathology* 58, 1164–1167.

Briggs, M.H. (1981) *Vitamins in Human Biology and Medicine*. CRC Press, Boca Raton, Florida.

Buitelaar, K. (1989) Proefstation voor Tuinbouw onder gla, Nalldwijk, Netherlands. *Groenten en Fruit* 45, 41.

Burton, G.W. and Ingold, K.U. (1984) Beta-carotene: An unusual type of lipid antioxidant. *Science* 224, 569–573.

Bush, W.A. (1936) Chile seed. *Journal of the American Chemistry Society* 58, 1821.

Buttery, R.G., Seifert, R.M., Guadagni, D.G. and Ling, L.C. (1969) Characterization of some volatile constituents of bell peppers. *Journal of Agriculture, Food and Chemistry* 17, 1322–1328.

Cantliffe, D.J. and Goodwin, P. (1975) Red color enhancement of pepper fruit by multiple applications of ethephon. *Journal of the American Society for Horticultural Science* 100(2), 157–161.

Capsicum & Eggplant Newsletter (CENL) (1994) Rules for gene nomenclature of *Capsicum*. 13, 13–14.

Carmichael, J.K. (1991) Treatment of herpes zoster and postherpetic neuralgia. *American Family Physician* 44(1), 203–210.

Castilla, N., Lorenzo, P., Montero, J.I., Fereres, E., Bretones, F., López-Gálvez, J. and Pérez-Parra, J. (1989) Alternative greenhouses for mild winter climate areas of Spain preliminary report. *Acta Horticulturae* 245, 63.

Cavero, J., Gil Ortega, R. and Zaragoza, C. (1996) Clear plastic mulch improved seedling emergence of direct-seeded pepper. *HortScience* 31(1), 70–73.

Chen, S.L. and Gutmanis, F. (1968) Auto-oxidation of extractable color pigments in chili pepper with special reference to ethoxyquin treatment. *Journal of Food Science* 33, 274–280.

Chen, P. and Lott, J.N. (1992) Studies of *Capsicum annuum* seeds: structure, storage reserves, and mineral nutrients. *Canadian Journal of Botany* 70, 518–529.

Cheng, S.S. (1989) The use of *Capsicum chinense* as sweet pepper cultivars and source for gene transfer. In: Griggs, T.D. and McLean, B.T. (eds) *Tomato and Pepper Production in the Tropics*. Asian Vegetable Research & Development Center, Taiwan, pp. 55–62.

Chile Pepper Institute (CPI) (1997/1998) Biocontrol popular in the chile greenhouse. *New Mexico State University Chile Pepper Institute Newsletter* 6, 1–2.

Chile Pepper Institute (CPI) (1997) Introducing Peruvian aji chiles. *Chile Pepper Institute Newsletter* 6 (3), 1–4.

Cichewicz, R.H. and Thorpe, P.A. (1996) The antimicrobial properties of chile peppers (*Capsicum* species) and their uses in Mayan medicine. *Journal of Ethnopharmacology* 52, 61–70.

Cochran, H.L. (1932) Factors affecting flowering and fruit-setting in the pepper. *American Society for Horticultural Science Proceedings* 29, 434–437.

Cochran, H.L. (1938) A morphological study of flower and seed development in pepper. *Journal of Agricultural Research* 56, 395–417.

Cochran, H.L. (1974) Effect of seed size on uniformity of pimiento transplants (*Capsicum annuum* L.) at harvest time. *Journal of the American Society for Horticultural Science* 99, 234–235.

Cochran, H.L. and Dempsey, A.H. (1966) Stigma structure and period of receptivity in pimientos (*Capsicum frutescens* L.). *American Society for Horticultural Science Proceedings* 88, 454–457.

Cohen, S. and Marco, S. (1973) Reducing the spread of aphid-transmitted viruses in peppers by trapping the aphids on sticky yellow polyethylene sheets. *Phytopathology* 63, 1207–1209.

Colditz, G.A. (1987) Beta-carotene and cancer. In: Quebedeaux, B. and Bliss, F.A. (eds) *Horticulture and Human Health, Contributions of Fruits and Vegetables*. ASHS Symposium Series No. 1. Prentice-Hall, New Jersey, pp.150–159.

Collins, M.D. and Bosland, P.W. (1994) Rare and novel capsaicinoid profiles in *Capsicum*. *Capsicum and Eggplant Newsletter* 13, 48–51.

Collins, M.D., Wasmund, L.M. and Bosland, P.W. (1995) Improved method for quantifying capsaicinoids in *Capsicum* using high-performance liquid chromatography. *HortScience* 30(1), 137–139.

Conicella, C., Saccardo, A.E. and Saccardo, F. (1990) Cytogenetic and isozyme studies of wild and cultivated *Capsicum annuum*. *Genome* 33, 279–282.

Conrad, R.S., Sundstrom, F.S. and Wilson, P.W. (1987) Evaluation of two methods of pepper fruit color determination. *HortScience* 22, 608–609.

Cornillon, P. and Palloix, A. (1997) Influence of sodium chloride on the growth and mineral nutrition of pepper cultivars. *Journal of Plant Nutrition* 20, 1085–1094.

Cotter, D.J. (1980) A review on the studies of chile. *New Mexico State University Bulletin* 673.

Cotter, D.J. (1986) Preplant broadcast and banded phosphate affects on growth and yield of long green mildly pungency chile. *New Mexico State University Research Report* 580.

Cotter, D.J. and Dickerson, G.W. (1984) Delayed harvest reduces yield of dry red chile in southern New Mexico. *HortScience* 19, 692–694.

Curtis, L.C. and Scarchuk, J. (1948) Seedless peppers; a single Mendelian recessive character. *Journal of Heredity* 39, 159–160.

Daeschel, M.A., Fleming, H.P. and Pharr, D.M. (1990) Acidification of brined cherry peppers. *Journal of Food Science* 55, 186–192.

Dainello, F.J. and Heineman, R.R. (1986) Plant arrangements and seedling establishment techniques for long green chile pepper production. *Texas Agriculture Experimental Station Progress Report* 4369.

Dainello, F. and Heineman, R.R. (1987) Influence of polyethylene-covered trenches on yield of bell pepper. *HortScience* 22, 225–227.

Daood, H.G., Biacs, P.A., Kiss-Kutz, N., Hajdu, F. and Czinkotai, B. (1989) Lipid and antioxidant content of red pepper. In: Biacs, P.A., Gruiz, K. and Kremmer, T. (eds) *Biological Role of Plant Lipids*. Plenum Publishing, New York, pp. 491–494.

Daskalov, S. and Mihailov, L. (1988) A new method for hybrid seed production based on cytoplasm male sterility combined with a lethal gene and a female sterile pollenizer in *Capsicum annuum* L. *Theoretical and Applied Genetics* 76, 530–532.

Daskalov, S. and Poulos, J.M. (1994) Updated Capsicum gene list. *Capsicum and Eggplant Newsletter* 13, 15–26.

Deal, C.L., Schmitzer, T.J., Lipstein, E., Seibold, J.P., Stevens, R.M., Levy, M.D., Albert, D. and Renold, F. (1991) Treatment of arthritis with topical capsaicin: a double blind trial. *Clinical Therapeutics* 13(3), 383–395.

Decoteau, D.R., Kasperbauer, M.J. and Hunt, P.G. (1990) Bell pepper plant development over mulches of diverse colors. *HortScience* 25, 460–462.

de la Vega, G. (1609) *Royal commentaries of the Incas*, trans. Livermore, H.V. University of Texas Press, Austin, Texas.

Deli, J. and Tiessen, H. (1969) Interaction of temperature and light intensity on flowering of *Capsicum frutescens* var. grossum cv. 'California Wonder'. *Journal of the American Society for Horticultural Science* 94, 349–351.

Demers, D.A., Charbonneau, J. and Gosselin, A. (1991) Effects of supplementary lighting on the growth and productivity of greenhouse sweet pepper. *Canadian Journal of Plant Science* 71, 587–594.

De Vaulx, D., Chambonnet, R. and Pochard, E. (1981) Culture in vitro d'antheres de piment (*Capsicum annuum* L.): amelioration des taux d'obtention de plantes chez differents genotypes par des traitments a +35 degrees C. *Agronomie* 1, 859–864.

DeWitt, D. and Bosland, P.W. (1993) *The Pepper Garden*. Ten Speed Press, Berkeley, California.

DeWitt, D. and Bosland, P.W. (1996) *Peppers of the World*. Ten Speed Press, Berkeley, California.

Diaz, I., Moreno, P.W. and Power, J.B. (1988) Plant regeneration from protoplasts of *Capsicum annuum*. *Plant Cell Reports* 7, 210–212.

Dorland, R.E. and Went, F.W. (1947) Plant growth under controlled conditions: VIII, Growth and fruiting of the chili pepper (*Capsicum annuum*). *American Journal of Botany* 34(8), 393–401.

Dufault, R.J. and Wiggins, S.C. (1981) Response of sweet peppers to solar reflectors and reflective mulches. *HortScience* 16, 65–67.

Duke, J.A. and Vasquez, R. (1994) *Amazonian Ethnobotanical Dictionary*. CRC Press, Boca Raton, Florida.

Dunal, F.M. (1852) The Solanaceae. In: de Candolle, A.P. (ed.) *Prodromous Systematis Naturalis Regni Vegetablilis* vol. 13, Masson, Paris, pp. 411–429.

Durán, F.D. (1588) *Historia de las Indias de Nueva-España y Islas de Tierra Firme*; trans. by Heyden, D. (1964) as *The History of the Indies of New Spain*. The Civilization of the American Indian Series, vol. 210. University of Oklahoma Press, Norman, Oklahoma.

Egawa, X. and Tanaka, M. (1986) Chromosome structural differentiation between *Capsicum annuum* var. *annuum* and *C. chinense*. *Capsicum Newsletter* 5, 18–19.

Erickson, A.N. and Markhart, A.H. III (1997) Development and abortion of flowers in *Capsicum annuum* exposed to high temperatures. *HortTechnology* 7(3), 312–313.

Erwin, A.T. (1937) Anthesis and pollination of the *Capsicum*. *American Society for Horticultural Science Proceedings* 28, 309.

Esch, T.A. and Marshall, D.E. (1987) Trash removal from mechanically harvested peppers. *Transactions of ASAE, American Society of Agricultural Engineering* 30(4), 893–897.

Eshbaugh, W.H. (1963) The role of cultivation in the evolution of the genus *Capsicum*. *American Journal of Botany* 50, 633–634.

Fairchild, D. (1938) *The World was my Garden*. Charles Scribner's Sons, New York.

FAO (1997) *1996 Yearbook* Production Vol. 50, Statistics Series No. 135. FAO, Rome.

Fari, M. (1995) Impact of cell and tissue culture techniques on the breeding of *Capsicum*. *IXth Meeting Genetics Breeding Capsicum and Eggplant, Budapest, Hungary*. European Association for Plant Breeding, Budapest, Hungary, pp. 53–59.

Fari, M. and Czako, M. (1981) Relationship between position and morphogenetic response of pepper hypocotyl explants cultured *in vitro*. *Scientia Horticulturae* 15, 207–213.

FDA (1979) *Acidified Foods. Current Good Manufacturing Practice*. Fed. Reg. 44, 16230, FDA, Washington, DC.

Fehr, W.R. (1993) *Principles of Cultivar Development*. Macmillan Publishing Co., New York, USA.

Fieldhouse, D.J. and Sasser, M. (1975) Stimulation of pepper seed germination by sodium hypochlorite treatment. *HortScience* 10, 622.

Fletcher, J.T. (1992) Disease resistance in protected crops and mushrooms. *Euphytica* 63, 33–49.

Fletcher, J.T. (1994) Fusarium stem and fruit rot of sweet peppers in the glasshouse. *Plant Pathology* 43, 225–227.

Flora, L.F. and Heaton, E.K. (1979) Processing factors affecting acidification of canned pimiento peppers. *Journal of Food Science* 44, 1498–1500.

Flora, L.F., Heaton, E.K. and Shewfelt, A.L. (1978) Evaluation of factors influencing variability of acidified canned pimientos. *Journal of Food Science* 43, 415–419.

Franceschetti, U. (1971) Natural cross pollination in pepper (*Capsicum annuum* L.) *Proceedings of the Eucarpia Meeting on Genetics and Breeding of Capsicum*. University of Turin, Turin, Italy, pp. 346–353.

Fullilove, H.M., III and Futral, J.G. (1972) *A Mechanical Harvester for Pimiento Peppers*. ASAE Paper No. 72-148, ASAE, 2950 Niles Rd, St Joseph, MI 49085-9659, USA.

Garcia, F. (1908) *Chile Culture*. New Mexico College of Agriculture and Mechanic Arts, Las Cruces, Bulletin No. 67.

Garcia, F. (1921) Improved variety No. 9 of native chile. *New Mexico State University, Agricultural Experiment Station Bulletin* 124.

Gerard, C.J. and Chambers, G. (1967) Effect of reflective coatings on soil temperatures, soil moisture, and the establishment of fall bell peppers. *Agronomy Journal* 59, 293–297.

Gerson, R. and Homna, S. (1978) Emergence response of the pepper at low soil temperature. *Euphytica* 27, 151–156.

Ghate, S.R. and Phatak, S.C. (1982) Performance of tomato and pepper seeds germinated before planting. *Journal of the American Society for Horticultural Science* 107, 908–911.

Ghatnekar, S.D. and Kulkarni, A.R. (1978) Studies on pollen storage in *Capsicum frutescens* and *Ricinus communis*. *Journal of Palynology* 14(2), 150–157.

Gill, H.S., Thakur, P.C. and Thakur, E.C. (1974) Effect of nitrogen and phosphorus application on seed yield of sweet pepper (*Capsicum annuum* L.). *Indian Journal of Horticulture* 31, 74–78.

Gillin, J. (1936) *The Barama River Caribs of British Guiana*. Papers of the Peabody Museum of American Archaeology and Ethnology, Vol. 14: No. 2. Harvard University, Cambridge, Massachusetts.

Goldberg, N. (1995) *Chile Pepper Diseases*. Agricultural Experiment Station Circular 549, New Mexico State University, Las Cruces.

Gosselin, A. and Trudel, M.J. (1986) Root-zone temperature effects on pepper. *Journal of the American Society for Horticultural Science* 111, 220–224.

Govindarajan, V.S. (1988) Capsicum: Production, technology, chemistry, and quality. *CRC Critical Reviews in Food Science Nutrition* 25(3), 185–282.

Green, S.K. and Kim, J.S. (1991) *Characteristics and Control of Viruses Infecting Peppers: a Literature Review*, Technical Bulletin, No. 18. Asian Vegetable Research and Development Center, Taipei, Taiwan.

Greenleaf, W.H. (1986) Pepper breeding. In: Bassett, M.J. (ed.) *Breeding Vegetable Crops*. AVI Publishing Co., Westport, Connecticut, pp. 69–127.

Gunay, A.L. and Rao, P.S. (1978) *In vitro* plant regeneration from hypocotyl and cotyledon explants of red pepper. *Plant Science* 11, 365–372.

Haas, J.H., Bar-Tal, A., Bar-Yosef, B. and Krikun, J. (1986) Nutrient availability effects on vesicular-arbuscular mycorrhizal bell pepper (*Capsicum annuum*) seedlings and transplants. *Annals of Applied Biology* 108, 171–179.

Haile, Y. and Zewdie, Y. (1989) Hot pepper and tomato production and research in Ethiopia. In: Green, S.K., Griggs, T.D. and McLean, B.T. (eds) *Tomato and Pepper Production in the Tropics*. AVRDC, Taiwan, pp. 442–451.

Harkay-Vinkler, M. (1974) Storage experiments with raw material of seasoning paprika, with particular reference to the red pigment components. *Acta Alimentaria Academiae Scientarium Hungaricae* 3, 239–249.

Harn, C., Kim, M.Z., Choi, K.T. and Lee, Y.I. (1974) Production of haploid callus and embryoid from the cultured anther of *Capsicum annuum*. *Sabrao Journal* 7, 71–77.

Harrington, J.F. (1960) Germination of seeds from carrot, lettuce, and pepper plants grown under severe nutrient deficiencies. *Hilgardia* 20, 219–255.

Harrison, M.K. and Harris, N.D. (1985) Effects of processing treatments on recovery of capsaicin in jalapeño peppers. *Journal of Food Science* 50, 1764–1765.

Hartwell, J.L. (1971) Plants used against cancer. A Survey. *Lloydia* 34, 204–244.

Hartz, T.K., Bogle, C.R. and Villalon, B. (1985) Response of pepper and muskmelon to row solarization. *HortScience* 20, 699–701.

Hartz, T.K., LeStrange, M. and May, D.M. (1993) Nitrogen requirements of drip-irrigated peppers. *HortScience* 28, 1097–1099.

Harvell, K.P. and Bosland, P.W. (1997) The environment produces a significant effect on pungency of chiles. *HortScience* 32, 1292.

Haymon, L.W. and Aurand, L.W. (1971) Volatile constituents of Tabasco peppers. *Journal of Agriculture, Food Chemistry* 19, 1131–1134.

Haytowitz, D.B. and Matthews, R.H. (1984) *Composition of Foods, Vegetables, and Vegetable Products – Raw, Processed Prepared*. USDA Agricultural Handbook 8–11, Washington, DC.

Heiser, C.B. (1976) Peppers *Capsicum* (Solanaceae). In: Simmonds, N.W. (ed.) *Evolution of Crop Plants*. Longman Press, London, pp. 265–268.

Heiser, C.B. and Smith, P.G. (1953) The cultivated *Capsicum* peppers. *Economic Botany* 7(3), 214–217.

Heiser, C.B. and Smith, P.G. (1958) New species of *Capsicum* from South America. *Brittonia* 10, 194–201.

Hirose, T. (1957) Studies on the pollination of red pepper. I. Flowering and the germinability of the pollen. *Plant Breeding Abstracts* 585/1959.

Horner, H.T. Jr and Rogers, M.A. (1974) A comparative light and electron microscopic study of microsporogenesis in male-fertile and cytoplasmic male-sterile pepper (*Capsicum annuum* L.). *Canadian Journal of Botany* 52(3), 435–441.

Howard, L.R., Smith, R.T., Wagner, A.B., Villalon, B. and Burns, E.E. (1994) Provitamin A and ascorbic acid content of fresh pepper cultivars (*Capsicum annuum*) and processed jalapeños. *Journal of Food Science* 59, 362–365.

Huffman, V.L., Schadel, E.R., Villalon, B. and Burns, E.E. (1978) Volatile components and pungency in fresh and processing jalapeño peppers. *Journal of Food Science* 43, 1809.

Hunziker, A.T. (1979) South America Solanaceae: a synoptic survey. In: Hawkes, J.G., Lester, R.N. and Skelding, A.D. (eds) *The Biology and Taxonomy of the Solanaceae*. Academic Press, London, pp. 49–85.

IBPGR (1983) *Genetic Resources of Capsicum: A Global Plan of Action*. International Board for Plant Genetic Resources, Rome.

Inai, S., Ishikawa, K., Nunomura, O. and Ikehashi, H. (1993) Genetics analysis of stunted growth by nuclear-cytoplasmic interaction hybrids of *Capsicum* by using RAPD markers. *Theoretical and Applied Genetics* 87, 416–422.

Irish, H.C. (1898) A revision of the genus *Capsicum* with especial reference to garden varieties. *Annual Report of the Missouri Botanical Garden, St Louis* 9, 53–110.

Irwin, C.C. and Price, H.C. (1981) Sensitivity of pregerminated pepper seed to low temperatures. *Journal of the American Society for Horticultural Science* 106, 187–189.

Itoh, T., Jeong, T.M., Hirano, Y., Tamura, T. and Matsumoto, T. (1977) Occurrence of lanosterol and lanostenol in seed of red pepper (*Capsicum annuum*). *Steroids* 29, 569–577.

Itoh, T., Ishii, T., Tamura, T. and Matsumoto, T. (1978) Four new and other 4α-methylsterols in the seeds of Solanaceae. *Phytochemistry* 17, 971–977.

Itoh, T., Sakurai, S., Tamura, T. and Matsumoto, T. (1979) Occurrence of 24(E)-ethylidene sterols in two solanaceae seed oils and rice bran oil. *Lipids* 15, 22–25.

Jachimoiwiez, Th. (1941) Vitamin C in paprika. *Biochemische Zeitschrift* 307, 387–399.

Jacquin, N.J. (1776) *Hortus Botanicus Vindobonensis*. Vienna.

John-Shang, K., Yu-Ying, W., Nan-Fan, C. and Shu-Jong, K. (1973) Investigations on the anther culture in vitro of *Nicotiana tabacum* L. and *Capsicum annuum* L. *Acta Botanica Sinica* 15, 47–52.

Johnson, M. (1977) *Canning Green Chile*. New Mexico Cooperative Extension Guide E-308, Las Cruces.

Jones, D.R., Unwin, C.H. and Ward, E.W.B. (1975) Capsidiol induction in pepper fruit during interactions with *Phytophthora capsici* and *Monilinia fructicola*. *Phytopathology* 65, 1417–1419.

Jurenitsch, J. (1981) Scharfstoffzusammensetzung in Fruchten definierter Capsicum-Sippen-Konsequenzen fur Qualitatsforderunzen und Taxonomische Aspekte. *Scientia Pharmaceutica* 49, 321.

Kaga, H., Goto, K., Fukuda, T. and Orito, K. (1992) An alternative synthesis of the capsaicinoids. *Bioscience, Biotechnology and Biochemistry* 56, 946–948.

Kahn, B.A., Motes, J.E. and Maness, N.O. (1997) Use of ethephon as a controlled abscission agent on paprika pepper. *HortScience* 32, 251–255.

Kanchan, S.D. (1973) Effect of hardening on emergence and seedling growth in sweet pepper (var. California Wonder). *Current Science* 42(21), 762–763.

Kanner, J., Harel, S., Palevitch, D. and Ben-gera, I. (1977) Color retention in sweet paprika powder as affected by moisture contents and ripening stage. *Journal of Food Technology* 12, 59–64.

Kanner, J., Harel, S. and Mendel, H. (1979) Content and stability of α-tocopherol in fresh and dehydrated pepper fruits (*Capsicum annuum* L.). *Journal of Agriculture, Food and Chemistry* 27, 1316–1318.

Kano, K., Fujimura, T., Hirose, T. and Tsukamoto, Y. (1957) Studies on the thickening growth of garden fruits. I. On the cushaw, eggplant, and pepper. *Kyoto University Research Institute Food Science Memorandum* 12, 45–90.

Keller, U., Flath, R.A., Mon, T.R. and Terranishi, R. (1981) Volatiles from red pepper (*Capsicum* spp.). In: Terranishi, R. and Barrera-Berirtz, H. (eds) *Quality of Selected Fruit and Vegetables of North America*, ACS Symposium Series 170. American Chemical Society, Washington, DC, Chapter 12.

Keng, J.C.W., Scott, T.W. and Lugo-Lopez, M.A. (1979) Fertilizer management with drip irrigation in an Oxisol. *Agronomy Journal* 71, 971–980.

Khademi, M. and Khosh-Khui, M. (1977) Effect of growth regulators on branching, flowering and fruit development of ornamental pepper (*Capsicum annum* L.). *Journal of the American Society of Horticultural Sciences* 102, 796–798.

Khah, E.M. and Passam, H.C. (1992) Sodium hypochlorite concentration, temperature, and seed age influence germination of sweet pepper. *HortScience* 27, 821–823.

Kinsella, J.E. (1971) Composition of the lipids of cucumber and peppers. *Journal of Food Science* 36, 865–866.

Knavel, D.E. and Kemp, T.R. (1983) Ethephon and CPTA on color development in bell pepper fruits. *HortScience* 8, 403–404.

Krajewska, A.M. and Powers, J.J. (1988) Sensory properties of naturally occurring capsaicinoids. *Journal of Food Science* 53, 902–905.

Laborde, J.A. and Pozo, O. (1982) *Present and Past of the Chile Pepper in Mexico*. National Institute for Agricultural Research Special Publication No. 85, Mexico City (in Spanish).

Lai, M. (1976) Bacterial canker of bell pepper caused by *Corynebacterium michiganense*. *Plant Disease Reporter* 60, 339–342.

Lanteri, S. and Pickersgill, B. (1993) Chromosomal structural changes in *Capsicum annuum* L. and *C. chinense* Jacq. *Euphytica* 67, 155–160.

Lantz, E.M. (1943) The carotene and ascorbic acid content of peppers. *New Mexico Agricultural Experiment Station Bulletin* 306.

Lantz, E.M. (1946) Effects of canning and drying on the carotene and ascorbic acid content of chiles. *New Mexico Experimental Station Bulletin* 327.

Lease, J.G. and Lease, E.J. (1956) Factors affecting the retention of red color of red peppers. *Food Technology* 10, 368–373.

Lee, D.S., Chung, S.K. and Yam, K.L. (1992) Carotenoid loss in dried red pepper products. *International Journal of Food Science and Technology* 27, 179–185.

Lee, R.D. and Schroeder, J. (1995) *Weed Management in Chile*. New Mexico State University Circular 548, Las Cruces.

Lee, Y., Howard, L.R. and Villalon, B. (1995) Flavonoids and antioxidant activity of fresh pepper (*Capsicum annuum*) cultivars. *Journal of Food Science* 60, 473–476.

Lefebvre, V., Palloix, A. and Rives, M. (1993) Nuclear RFLP between pepper cultivars (*Capsicum annuum* L.). *Euphytica* 71, 189–199.

Lefebvre, V.A., Caranta, C., Pflieger, S., Moury, B., Daubeze, A.M., Blattes, A., Ferriere, C., Phaly, T., Nemouchi, G., Fufinattok, A. and Palloix, A. (1997) Updated intraspecific maps of pepper. *Capsicum and Eggplant Newsletter* 16, 35–41.

Lenker, D.H. and Nascimento, D.F. (1982) Mechanical harvesting and cleaning of chile peppers. *Transactions of the ASAE* 25, 42–46.

Leskovar, D.I. and Cantliffe, D.J. (1993) Comparison of plant establishment method, transplant, or direct seeding on growth and yield of bell pepper. *Journal of the American Society for Horticultural Science* 118, 17–22.

Linnaeus, C. (1737) *Genera Plantarum*. Stockholm.

Linnaeus, C. (1753) *Species Plantarum*. Stockholm.

Linnaeus, C. (1767) *Mantissa Plantarum*. Stockholm.

Lippert, L.F., Bergh, B.O. and Smith, P.G. (1965) Gene list for the pepper. *Journal of Heredity* 56, 30–34.

Loaiza-Figueroa, F., Ritland, K., Cancino, J. and Tanksley, S.D. (1989) Patterns of genetic variation of the genus *Capsicum* (Solanaceae) in Mexico. *Plant System Evolution* 165, 159–188.

Locascio, S.J., Fiskell, J.G.A. and Martin, F.G. (1981) Response of bell pepper to nitrogen source. *Journal of the American Society for Horticultural Science* 106, 628–632.

Lockwood, D. and Vines, H.M. (1972) Red color enhancement of pimiento peppers with (2-chloroethyl)phosphonic acid. *Journal of the American Society for Horticultural Science* 97, 192–197.

Loebenstein, G., Alper, M. and Levy, S. (1970) Field tests with oil sprays for the prevention of aphid-spread viruses in peppers. *Phytopathology* 60, 212–215.

Lopez, F. and Silvas, R.J. (1979) Study of planting distances between rows and plants in sweet peppers. *Proceedings of the American Society for Horticultural Science of the Tropical Region* 23, 207–210.

Lorenz, O.A. and Tyler, K.B. (1983). Plant tissue analysis of vegetable crops. In: Reisenauer, H.M. (ed.) *Soil and Plant Tissue Testing in California*. Bulletin 1897, University of California, pp. 24–29.

Lownds, N.K., Banaras, M. and Bosland, P.W. (1993) Relationship between postharvest water loss and physical properties of pepper fruit (*Capsicum annuum* L.). *HortScience* 28, 1182–1184.

Lownds, N.K., Banaras, M. and Bosland, P.W. (1994) Postharvest water loss and storage quality of nine pepper capsicum cultivars. *HortScience* 29, 191–193.

Lunin, J., Gallatin, M.H. and Batchelder, A.R. (1963) Saline irrigation of several vegetable crops at various growth stages. Effect on yields. *Agronomy Journal* 55, 107–114.

Lyons, J.M. and Lippert, L.F. (1966) Characterization of fatty acids from roots and shoot lipids of *Capsicum* species. *Lipids* 1(2), 136.

MacNeish, R.S. (1964) Ancient Mesoamerican civilization. *Science* 143, 531–537.

Marcelis, L.F.M. and Baan Hofman-Eijer, L.R. (1997) Effects of seed number on competition and dominance among fruits in *Capsicum annuum* L. *Annals of Botany* 79, 687–693.

Marshall, D.E. (1981) Performance of an open-helix mechanical harvester in processing peppers. ASAE Paper and Microfiche No. 81-1069.

Marshall, D.E. (1997) Designing a pepper for mechanical harvesting. *Capsicum and Eggplant Newsletter* 16, 15–27.

Marshall, D.E. and Brook, R.C. (1997) Reducing bell pepper bruising during postharvest handling. *HortScience* 32, 522.

Marshall, D.E. and Esch, T.A. (1986) Recovery and damage of mechanically harvested peppers. *Transactions of the ASAE* 29(2), 398–401.

Martinez, M. and Aljaro, A. (1987) Agronomic evaluation of the osmotic conditioning on sweet pepper seeds (*Capsicum annuum* L.) II. Effects on emergence and development of seedlings. *Agricultura Tecnica (Santiago)* 47, 321–325.

Matsumoto, T., Osawa, Y. and Itoh, T. (1983) 4α-Methyl-5α-cholets-8(14) en-3β-ol from the seeds of *Capsicum annuum. Phytochemistry* 22, 2621–2622.

Matus Z., Deli, J. and Szabolcs, J.J. (1991) Carotenoid composition of yellow pepper during ripening—isolation of B-cryptoxanthin 5,6-epoxide. *Journal of Agricultural and Food Chemistry* 39, 1907–1914.

Maynard, D.N. and Hochmuth, G.J. (1997) *Knott's Handbook for Vegetable Growers*, 4th edn. Wiley, New York.

Maynard, D.N., Lackman, W.H., Check, R.M. and Vernell, H.F. (1962) The influence of nitrogen levels on flowering and fruit set of peppers. *Proceedings of the American Society for Horticultural Science* 81, 385–389.

McBryde, F.W. (1933) *Sololá: A Guatemalan Town and Cakchiquel Market-center*. Middle America Research Series Publication 5, Pamphlet 3. Tulane University, New Orleans, pp. 45–152.

McBryde, F.W. (1945) *Cultural and Historical Geography of Southwest Guatemala*. Institute of Social Anthropology, Publication No.4. Smithsonian Institution, Washington, DC.

McGrady, J.J. and Cotter, D.J. (1984) Anticrustant effects on chile pepper stand and yield. *HortScience* 19, 408–409.

McKee, L. (1998a) Chile skins as a fiber source. *NMSU Chile Conference*, Las Cruces, New Mexico.

McKee, L. (1998b) Can chile for year-round flavor. *The Chile Pepper Institute Newsletter* 7(1), 4–5.

McLeod, M.J., Eshbaugh, W.H. and Guttman, S.I. (1979) An electrophoretic study of *Capsicum* (Solanaceae): The purple flower taxa. *Bulletin of Torrey Botanical Club* 106, 326–333.

McLeod, M.J., Guttman, S.I. and Eshbaugh, W.H. (1983) An electrophoretic study of evolution in *Capsicum* (Solanaceae). *Evolution* 37, 563–574.

Mejia, L.A., Hudson, E., Gonzalez de Mejia, E. and Vazquez, F. (1988) Carotenoid content and Vitamin A activity of some common cultivars of Mexican peppers (*Capsicum annuum*) as determined by HPLC. *Journal of Food Science* 53(5), 1448–1451.

Miller, C.H. (1961) Some effects of different levels of five nutrient elements on bell peppers. *Proceedings of the American Society for Horticultural Science* 77, 440–448.

Miller, C.H., McCollum, R.E. and Claimon, S. (1979) Relationships between growth of bell peppers (*Capsicum annuum* L.) and nutrient accumulation during ontogeny and field environments. *Journal of the American Society for Horticultural Science* 104, 852–857.

Morales-Payan, J.P., Santos, B.M., Stall, W.M. and Beiwick, T.A. (1997) Effects of purple nutsedge (*Cyperus rotundus*) on tomato (*Lycopersicon esculentum*) and bell pepper (*Capsicum annuum*) vegetative growth and fruit yield. *Weed Technology* 11, 672–676.

Morrison, J.E., Jr, Milbocker, D.C., Atkinson, W.O. and Smiley, J.H. (1973) Transplanter modification and survival of transplants under no-tillage conditions. *HortScience* 8, 483–485.

Morrison, R. (1699) *Plantarum Historiae Universalis Oxoniensis*. Paulum & Isaacum Vaillant, London.

Moscone, E.A. (1990) Chromosome studies on *Capsicum* (Solanaceae) I. Karyotype analysis in *C. chacoense*. *Brittonia* 42, 147–154.

Moscone, E.A., Lambrou, M., Hunziker, A.T. and Ehrendorfer, F. (1993) Geisma C-banded karyotypes in *Capsicum* (Solanaceae). *Plant System Evolution* 186, 213–229.

Moscone, E.A., Loidi, J., Ehrendorfer, F. and Hunziker, A.T. (1995) Analysis of active nucleolus organizing regions in *Capsicum* (Solanaceae) by silver staining. *American Journal of Botany* 82, 276–287.

Motsenbocker, C. (1996) In-row plant spacing affects growth and yield of pepperoncini pepper. *HortScience* 31, 198–200.

Mozafar, A. (1994) *Plant Vitamins: Agronomic, Physiological and Nutritional Aspects*. CRC Press, Boca Raton, Florida.

Muhyi, R. and Bosland, P.W. (1995) Evaluation of *Capsicum* germplasm for sources of resistance to *Rhizoctonia solani*. *HortScience* 30, 341–342.

Nabhan, G.P. (1985) *Gathering the Desert*. The University of Arizona Press, Tucson, Arizona, pp.123–124.

Nagarathnam, A.K. and Rajamani, T.S. (1963) Germination studies in chillies, *Capsicum annuum* Linn. *Madras Agriculture Journal* 50, 200.

National Academy of Science (1982) *Diet, Nutrition and Cancer*. In: Grobstein, C. (Chairman). A report by the Committee on Diet, Nutrition, and Cancer, Assembly of Life Sciences. National Academy Press, Washington, DC.

National Research Council (NRC), Food and Dietary Board (1989) *Recommended Dietary Allowances*. National Academy of Sciences, Washington DC.

Negulesco, J.A., Lohse, C.L., Hrabovsky, E.E., Boggs, M.T. and Davis, D.H. (1989) Dihydrocapsaicin protects against serum hyperlipidemia in guinea pigs fed a cholesterol-enriched diet. *Artery* 16, 174–188.

Nervo, G., Ferrari, V. and Caporali, E. (1995) Evaluation of anther culture derived plants of pepper. *IXth Meeting Genetics Breeding Capsicum and Eggplant*. Budapest, Hungary. European Association for Plant Breeding.

New Mexico Crop Improvement Association (NMCIA) (1992) *Official Seed Certification Handbook*. New Mexico State University, Las Cruces, New Mexico, p. 38.

Nisperos-Carriedos, M.O., Buslig, B.S. and Shaw, P.E. (1992) Simultaneous detection of dehydroascorbic, ascorbic, and some organic acids in fruits and vegetables by HPLC. *Journal of Agriculture and Food Chemistry* 40, 1127–1130.

Nitzany, F.E. (1966) Tests for the control of field spread of pepper viruses by oil spray. *Plant Disease Reporter* 50, 158–160.

Novak, F., Betlach, J. and Dubovsky, J. (1971) Cytoplasm male sterility in sweet pepper (*Capsicum annuum* L.). *Zeitschrift Pflanzenzuchtg* 65, 129–140.

O'Dell, C.R., Scyphers, E. and Conner, C. (1979) Early season production of bell peppers in row tunnels. *The Vegetable Growers News* 34, 2.

Odland, M.L. and Porter, A.M. (1941) A study of natural crossing in peppers (*Capsicum frutescens*). *American Society for Horticultural Science Proceedings* 38, 585–588.

O'Sullivan, J. and Bouw, W.J. (1984) Pepper seed treatment for low-temperature germination. *Canadian Journal of Plant Science* 64, 387–393.

Osuna-Garcia, J.A. (1996) Red color retention during paprika (*Capsicum annuum* L.) storage as affected by natural antioxidants and moisture content. PhD dissertation, New Mexico State University, Las Cruces, New Mexico, USA.

Osuna-Garcia, J.A. and Wall, M.M. (1998) Prestorage moisture content affects color loss of ground paprika (*Capsicum annuum* L.) under storage. *Journal of Food Quality* 21, 251–259.

Osuna-Garcia, J.A., Wall, M.M. and Waddell, C.A. (1998) Endogenous levels of tocophenols and ascorbic acid during fruit ripening of New Mexican-type chile (*Capsicum annum* L.) cultivars. *Journal of Agricultural and Food Chemistry* 46, 5093–5096.

Ozaki, H.Y. and Hamilton, M.G. (1954) Bronzing and yield of peppers as influenced by varying levels of nitrogen, phosphorus, and potassium fertilization. *Soil Crop Science Society Fla. Proceedings* 14, 185–189.

Palloix, A., Daubeze, A.M., Phaly, T. and Pochard, E. (1990a) Breeding transgressive lines of pepper for resistance to *Phytophtora capsici* in a recurrent selection system. *Euphytica* 51, 141–150.

Palloix, A., Pochard, E., Phaly, T. and Daubeze, A.M. (1990b) Recurrent selection for resistance to *Verticillium dahliae* in pepper. *Euphytica* 47, 79–89.

Panpruik, P., McCaslin, B.D. and Wierenga, P.J. (1982) Effects of nitrogen and phosphorus fertilizer on yield and leaf content of trickle irrigated chile peppers. *New Mexico Agricultural Experiment Station Research Report* 480.

Paterson, A.H., Tanksley, S.D. and Sorrells, M.E. (1991) DNA markers in plant improvement. *Advances in Agronomy* 46, 39–91.

Payero, J.O., Bhangoo, M.S. and Steiner, J.J. (1990) Nitrogen fertilizer management practices to enhance seed production by 'Anaheim Chili' peppers. *Journal of the American Society for Horticultural Science* 115, 245–251.

Perry, K.B., Bonanno, A.R. and Monks, D.W. (1992) Two putative cryoprotectants do not provide frost and freeze protection in tomato and pepper. *HortScience* 27, 26–27.

Peterson, P.A. (1958) Cytoplasmically inherited male sterility in *Capsicum*. *American Naturalist* 863, 111–119.

Peto, R. (1983) The marked differences between carotenoids and retinoids: methodological implications for biochemical epidemiology. *Cancer Surveys* 2, 327–340.

Phillips, G.C. and Hubstenberger, J.F. (1985) Organogenesis in pepper tissue cultures. *Plant Cell, Tissue and Organ Culture* 4, 261–269.

Pickersgill, B. (1971) Relationships between weedy and cultivated forms in some species of chili peppers (Genus *Capsicum*). *Evolution* 25, 683–691.

Pickersgill, B. (1997) Genetic resources and breeding of *Capsicum* spp. *Euphytica* 96, 129–133.

Pitt, G.A.J. (1979) Vitamin A deficiency and excess. In: Geoffrey, T.T. (ed.) *The Importance of Vitamins to Human Health*. University Park Press, Baltimore.

Porter, W.C. and Etzel, W.W. (1982) Effects of aluminum-painted and black polyethylene mulches on bell pepper, *Capsicum annuum* L. *HortScience* 17, 942–943.

Portree, J. (ed.) (1996) *Greenhouse Vegetable Production Guide for Commercial Growers*. Ministry of Agriculture, Fisheries, and Food, Province of British Columbia.

Posselius, J.H., Jr and Valco, T.D. (1985) Pepper harvester research in Texas. *Papers of the American Society Agriculture Engineers Microfiche Collection* (fiche no. 85-1064).

Powers, J.J., Morse, R.E., Sane, R.H. and Mills, W.C. (1950) Acidification and calcium-firming of canned pimientos. *Food Technology* 4, 485–488.

Powers, J.J., Pratt, D.E., Downing, D.L. and Powers, I.T. (1961) Effect of acid level, calcium salts, monosodium glutamate and sugar on canned pimientos. *Food Technology* 15, 67–74.

Prince, J.P., Loaiza-Figueroa, F. and Tanklsey, S.D. (1992) Restriction fragment length polymorphism and genetic distance among Mexican accessions of *Capsicum*. *Genome* 35, 726–732.

Prince, J.P., Pochard, E. and Tanksley, S.D. (1993) Construction of a molecular linkage map of pepper and a comparison of synteny with tomato. *Genome* 36, 404–417.

Prince, J.P., Lackney, V.K., Angeles, C., Blauth, J.R. and Kyle, M.M. (1994) A survey of DNA polymorphism within the genus *Capsicum* and the fingerprinting of pepper cultivars. *Genome* 38, 224–231.

Quagliotti, L. (1979) Floral biology of *Capsicum* and *Solanum melongena*. In: Hawkes, J.G., Lester, R.N. and Skelding, A.D. (eds) *The Biology and Taxonomy of the Solanaceae*. Academic Press, New York, pp. 399–419.

Randle, W.M. and Homna, S. (1980) Inheritance of low temperature emergence in *Capsicum baccatum* var. *pendulum*. *Euphytica* 29, 331–335.

Randle, W.M. and Homna, S. (1981) Dormancy in peppers. *Scientia Horticulturae* 14, 19–25.

Reddy, B.S. and Sarojini, G. (1987) Chemical and nutritional evaluation of chili (*Capsicum annuum*) seed oil. *Journal of the American Oil Chemists' Society* 64, 1419–1422.

Reeves, M.J. (1987) Re-evaluation of *Capsicum* color data. *Journal of Food Science* 52, 1047–1049.

Rivas, M., Sundstrom, F.J. and Edwards, R.I. (1984) Germination and crop development of hot pepper after seed priming. *HortScience* 19, 279–281.

Ruiz, H. and Pavon, J. (1790) *Flora Peruviana, et Chilensis*, etc., 2, 30–31; cited in Heiser and Smith (1958).

Rylski, I. and Halevy, A.H. (1972) Factors controlling the readiness to flower of buds along the main axis of pepper (*Capsicum annuum* L.). *Journal of the American Society for Horticultural Science* 97, 309–312.

Saccarod, F. and La Gioria, N. (1982) Translocation studies in *Capsicum annuum* L. *Capsicum Newsletter* 1, 14–15.

Sachs, M., Cantliffe, D.J. and Nell, T.A. (1981) Germination studies of clay-coated sweet pepper seeds. *Journal of the American Society for Horticultural Science* 106, 385–389.

Sachs, M., Cantliffe, D.J. and Nell, T.A. (1982) Germination behavior of sand-coated sweet pepper seed. *Journal of the American Society for Horticultural Science* 107, 412–416.

Sahagún, B. de (1590) *The General History of the Things of New Spain; Florentine Codex*; transl. Anderson, A.J.O. and Dibble, C.E. School of American Research (Santa Fe) Monograph No. 14. Santa Fe, New Mexico.

Sahin, F. and Miller, S.A. (1996) Characterization of Ohio strains of *Xanthomonas campestris* pv. *vesicatoria*, causal agent of bacterial spot of pepper. *Plant Diseases* 80, 773–778.

Saldana, G. and Meyer, R. (1981) Effects of added calcium on texture and quality of canned jalapeño peppers. *Journal of Food Science* 46, 1518–1520.

Sanders, D.C., Kirk, H.J. and Van der Brink, C. (1980) *Growing Degree Days in North Carolina*. North Carolina Agricultural Extension Service, Raleigh, AG-236.

Sapers, G.M., Carre, J., DiVito, A.M. and Panasiuk, O. (1980) Factors affecting the pH of home-canned peppers. *Journal of Food Science* 45, 726–729.

Sawhney, V.K. (1981) Abnormalities in pepper (*Capsicum annuum*) flowers induced by gibberellic acid. *Canadian Journal of Botany* 59, 8–15.

Saxena, P.K., Gill, R., Rashid, A. and Maheshwari, S.C. (1981) Isolation and culture of protoplasts of *Capsicum annum* L. and their regeneration into plants flowering in vitro. *Protoplasma* 108, 357–360.

Schoch, P.G. (1972) Effects of shading on structural characteristics of the leaf and yield of fruit in *Capsicum annuum* L. *Journal of the American Society for Horticultural Science* 97, 461–464.

Schultheis, J.R., Cantliffe, D.J., Bryan, H.H. and Stoffella, P.J. (1988a) Planting methods to improve stand establishment, uniformity, and earliness to flower in bell pepper. *Journal of the American Society for Horticultural Science* 113, 331–335.

Schultheis, J.R., Cantliffe, D.J., Bryan, H.H. and Stoffella, P.J. (1988b) Improvement of plant establishment in bell pepper with a gel mix planting medium. *Journal of the American Society for Horticultural Science* 113, 546–552.

Scoville, W.L. (1912) Note on Capsicum. *Journal of the American Pharmaceutical Association* 1, 453.

Setiamihardja, R. and Knavel, D.E. (1990) Association of pedicel length and diameter with fruit length and diameter and ease of fruit detachment in pepper. *Journal of the American Society for Horticultural Science* 115, 677–681.

Shah, J.J. and Patel, J.D. (1970) Morpho-histogenetic studies in vegetative and floral buds of brinjal and chili. *Phytomorphology* 20, 209–221.

Shearer, S.A. and Payne, F.A. (1990) Color and defect sorting of bell peppers using machine vision. *Transactions of the ASAE* 33(6), 2045–2050.

Shifriss, C. (1997) Male sterility in pepper (*Capsicum annuum* L.). *Euphytica* 93, 83–88.

Shifriss, C. and Frankel, R. (1969) A new male sterility gene in *Capsicum annuum* L. *Journal of the American Society for Horticultural Science* 94, 385–387.

Shifriss, C. and Frankel, R. (1971) New sources of cytoplasmic male sterility in cultivated peppers. *Journal of Heredity* 62, 254–256.

Shifriss, C. and Guri, A. (1979) Variation in stability of cytoplasmic-genic male sterility in *Capsicum annuum* L. *Journal of the American Society for Horticultural Science* 104, 94–96.

Shifriss, C. and Pilovsky, M. (1993) Digenic nature of male sterility in pepper (*Capsicum annuum* L.). *Euphytica* 67, 111–112.

Shifriss, C. and Rylsky, I. (1972) A male sterile (*ms-2*)gene in 'California Wonder' pepper (*Capsicum annuum* L.). *HortScience* 7, 36.

Sibi, M., Dumas De Vaulx, R. and Chambonnet, D. (1979) Obtention de plantes haploides par androgenese in vitro chez le Piment (*Capsicum annuum* L.). *Annales de l'amelioration des Plantes* 29, 583–606.

Sicuteri, F., Fanciullacci, M., Nicolodi, M., Geppetti, P., Fusco, B.M., Marabini, S., Alessandri, M. and Campagnolo, V. (1990) Substance P theory: a unique focus on the painful and painless phenomena of cluster headache. *Headache* 30, 69–79.

Silver, W.L. and Maruniak, J.A. (1981) Trigeminal chemoreception in the nasal and oral cavities. *Chemical Senses* 6, 295–305.

Simons, J.N. (1960) Effect of foliar sprays of cytovirin on susceptibility to and transmissibility of potato virus Y in pepper. *Phytopathology* 50, 109–111.

Smith, P.G. and Heiser, C.B. (1957) Taxonomy of *Capsicum sinense* Jacq. and the geographic distribution of the cultivated *Capsicum* species. *Bulletin of Torrey Botanical Club* 84, 413–420.

Smith, P.G., Villalon, B. and Villa, P.L. (1987) Horticultural classification of pepper grown in the United States. *HortScience* 22, 11–13.

Smith, P.T. and Cobb, B.G. (1991) Accelerated germination of pepper seed by priming with salt solutions and water. *HortScience* 26, 417–419.

Smith, R., Mullen, R. and Hartz, T. (1996) Epidemiology and control of pepper stip. *Proceedings of the National Pepper Conference, Naples, Florida.* University of Florida, pp.103–104.

Somos, A. (1984) *The Paprika.* Akademiai Kiado, Budapest.

Sosa-Coronel, J. and Motes, J.E. (1982) Effect of gibberellic acid and seed rates on pepper seed germination in aerated water columns. *Journal of the American Society for Horticultural Science* 107, 290–295.

Starman, T.W. (1993) Ornamental pepper growth and fruiting response to uniconazole depends on application time. *HortScience* 28, 917–919.

Staub, J. E., Serquen, F.C. and Gupta, M. (1996) Genetic markers, map construction, and their application in plant breeding. *HortScience* 31, 729–740.

Stoffella, P.J. and Bryan, H.H. (1988) Plant population influences growth and yields of bell pepper. *Journal of the American Society for Horticultural Science* 113, 835–839.

Stoffella, P.J., Lipucci Di Paola, M., Pardossi, A. and Tognoni, F. (1988) Root morphology and development of bell peppers. *HortScience* 23, 1074–1077.

Stroehlein, J.L. and Oebker, N.F. (1979) Effects of nitrogen and phosphorus on yields and tissue analysis of chili peppers. *Communications in Soil Science and Plant Analysis* 10, 551–563.

Stroup, W.H., Dickerson, R.W. and Johnston, M.R. (1985) Acid equilibrium in mushrooms, pearl onions, and cherry peppers. *Journal of Food Protection* 48, 590–594.

Sturtevant, E.L. (1919) *Sturtevant's Notes on Edible Plants.* Lyon, Albany, USA.

Sundstrom, F.J. and Edwards, R.L. (1989) Pepper seed respiration, germination, and seedling development following seed priming. *HortScience* 24, 343–345.

Sundstrom, F.J., Thomas, C.H., Edwards, R.L. and Baskin, G.R. (1984) Influence of nitrogen and plant spacing on mechanically harvested tabasco pepper. *Journal of the American Society for Horticultural Science* 109, 642–645.

Supran, M.K., Powers, J.J., Rao, P.V., Dornseifer, T.P. and King, P.H. (1966) Comparison of different organic acids for the acidification of canned pimientos. *Food Technology* 20, 117–122.

Sviribeley, J.L. and Szent-Györgyi, A. (1933) The chemical structure of vitamin C. *Biochemistry Journal* 27, 100–104.

Szallasi, A. and Blumberg, P.M. (1990) Resiniferatoxin and its analogs provide novel insights into the pharmacology of the vanilloid (capsaicin) receptor. *Life Sciences* 47, 1399–1408.

Tanksley, S.D. (1983) Molecular markers in plant breeding. *Plant Molecular Biology Reporter* 1, 3–8.

Tanksley, S.D. (1984a) High rates of cross-pollination in chile pepper. *HortScience* 19, 580–582.

Tanksley, S.D. (1984b) Linkage relationships and chromosomal locations of enzyme-coding genes in pepper, *Capsicum annuum. Chromosoma* 89, 352–360.

Tanksley, S.D. and Iglesias-Olivas, J. (1984) Inheritance and transfer of multiple-flower character from *Capsicum chinense* into *Capsicum annuum*. *Euphytica* 33, 769–777.

Tanksley, S.D., Bernatzky, R., Lapitan, N.L. and Prince, J.P. (1988) Conservation of gene repetoire but not gene order in pepper and tomato. *Proceedings of the National Academy of Sciences* 85, 6419–6423.

Thomas, J.R. and Oerther, G.F. (1972) Estimating nitrogen content of sweet pepper leaves by reflectance measurements. *Agronomy Journal* 64, 11–13.

Tong, N. (1998) Genetic relationships among *Capsicum* species. PhD thesis, New Mexico State University, Las Cruces, New Mexico, USA.

Tong, N. and Bosland, P.W. (1997) *Capsicum lanceolatum*, another 13 chromosome species. *Capsicum and Eggplant Newsletter* 16, 40–42.

Torabi, H. (1997) Identification of novel capsaicinoids. MS thesis, New Mexico State University, Las Cruces, New Mexico, USA.

Tourneforte, J.P. (1700) *Institutiones rei Herbariae*, 3 volumes. Paris.

Tripp, K.E. and Wien, H.C. (1989) Screening with ethephon for abscission resistance of flower buds in bell pepper. *HortScience* 24, 655–657.

Turner, A.D. and Wien, H.C. (1994) Dry matter assimilation and partioning in pepper cultivars differing in susceptibility to stress-induced bud and flower abscission. *Annals of Botany* 73, 617–622.

Ungs, W.D., Woodbridge, C.G. and Csizinszky, A.A. (1977) Screening peppers (*Capsicum annuum* L.) for resistance to curly top virus. *HortScience* 12, 161–162.

US Department of Health, Education, and Welfare (1968–1970) *Ten-State Nutritional Survey* IV. Biochemical Department HEW Pub. HSM 72-8132, Washington, DC.

Valencia, M.E., Jardines, R.P., Noriega, E., Cruz, R., Grijaliva, I. and Pena, C. (1983) The use of 24 hour recall data for nutrition surveys to determine food preference, availability, and food consumption baskets in populations. *Nutrition Reports International* 28, 815.

Van Blaricom, L.O. and Martin, J.A. (1951) Retarding the loss of red color in cayenne pepper with oil antioxidants. *Food Technology* 5, 337–339.

VanDerwerken, J.E. and Wilcox-Lee, D. (1988) Influence of plastic mulch and type and frequency of irrigation on growth and yield bell pepper. *HortScience* 23, 985–988.

Volcani, Z., Zutra, D. and Cohn, R. (1970) A new leaf and fruit spot disease of pepper caused by *Corynebacterium michiganense*. *Plant Disease Reporter* 54, 804–806.

Vos, J.G.M. and Sumarni, N. (1997) Integrated crop management of hot pepper (*Capsicum* spp.) under tropical lowland conditions: Effects of mulch on crop performance and production. *Journal of Horticultural Science* 72, 415–424.

Votava, E.J. and Bosland, P.W. (1996). Use of ladybugs to control aphids in *Capsicum* field isolation cages. *HortScience* 31(7), 1237.

Votava E.J. and Bosland, P. W. (1998) 'NuMex Piñata' jalapeño chile. *HortScience* 33, 350.

Wafer, L. (1699) *A New Voyage and Description of the Isthmus of America*; reprinted from the original edition. Burt Franklin, New York, p.107.

Wall, M.M. and Biles, C.L. (1993) Alternaria fruit rot of ripening chile peppers. *Phytopathology* 83, 324–328.

Wall, M.M. and Bosland, P.W. (1993) The shelf-life of chiles and chile containing products. In: Charalambous, G. (ed.) *Shelf Life Studies of Foods and Beverages*. Elsevier Science, Amsterdam, pp. 487–500.

Wall, M.M. and Bosland, P.W. (1998) Analytical methods for color and pungency of chiles (capsicums). In: Wetzel, D.L. and Charalambous, G. (eds) *Instrumental Methods in Food and Beverage Analysis.* Elsevier Science, Amsterdam, pp. 347–373.

Watkins, J.T. and Cantliffe, D.J. (1983a) Hormonal control of pepper seed germination. *HortScience* 18, 342–343.

Watkins, J.T. and Cantliffe, D.J. (1983b) Mechanical resistance of the seed coat and endosperm during germination of *Capsicum annuum* at low temperature. *Plant Physiology* 72, 146–150.

Watkins, J.T., Cantliffe, D.J., Huber, H.B. and Nell, T.A. (1985) Gibberellic acid stimulated degradation of endosperm in pepper. *Journal of the American Society of Horticultural Science* 110, 61–65.

Weisenfelder, A.E., Huffman, V.L., Villalon, B. and Burns, E.E. (1978) Quality and processing attributes of selected jalapeño pepper cultivars. *Journal of Food Science* 43, 885–887.

Welles, G.W.H. (1992) Integrated production systems for glasshouse horticulture. *Netherlands Journal of Agricultural Science* 40, 277–284.

Werner, D.J. and Honma, S. (1980) Inheritance of fruit detachment force in pepper. *Journal of the American Society of Horticultural Science* 105, 805–807.

Wiedenfeld, R.P. (1986) Rate, timing, and slow-release nitrogen fertilizers on bell peppers and muskmelon. *HortScience* 21, 233–235.

Wien, H.C., Tripp, K.E., Hernadez-Armenta, R. and Turner, A.D. (1989) Abscission of reproductive structures in pepper: causes, mechanisms, and control. In: Green, S.K, Griggs, T.D. and McLean, B.T. (eds) *Tomato and Pepper Production in the Tropics,* AVRDC, Taiwan, pp.150–165.

Willdenow, C.L. (1798) *Enumeratio Plantarum Horti regii botanici Berolinensis.* Germany.

Williams, J.G.K., Kubelik, A.R., Livak, K.J., Antoni, J.A. and Tingey, S.V. (1990) DNA polymorphisms amplified by arbitrary primers are useful as genetic markers. *Nucleic Acids Research* 18, 6531–6535.

Wimalasiri, P. and Wills, R.B.H. (1983) Simultaneous analysis of ascorbic acid and dehydroascorbic acid in fruits and vegetables by high-performance liquid chromatography. *Journal of Chromatography* 256, 368–371.

Wolfe, R.R. and Sandler, W.E. (1985) An algorithm for stem detection using digital image analysis. *Transactions of the ASAE* 28, 641–644.

Wolfe, R.R. and Swaminathan, M. (1986) Determining orientation and shape of bell peppers by machine vision. ASAE Paper No. 86-3045. American Society of Agriculture Engineers, St Joseph, Michigan, USA.

Yaklich, R.W. and Orzolek, M.D. (1977) Effect of polyethylene glycol-6000 on pepper seed. *HortScience* 12, 263–264.

Yaqub, C.M. and Smith, P.G. (1971) Nature and inheritance of self-incompatibility in *Capsicum pubescens* and *C. cardenasii. Hilgardia* 40, 459–470.

Zewdie-Tarekegn, Y. (1999) Genetic studies of capsaicinoids (pungency) in chile (Capsicum spp.). PhD dissertation, New Mexico State University, Las Cruces, USA.

Zhu, Y X., OuYang, W.J., Zhang, Y.F. and Chen, Z.L. (1996) Transgenic sweet pepper plants from *Agrobacterium* mediated transformation. *Plant Cell Reports* 16, 71–75.

Ziegler, R.G., Mason, T.J., Stemhagen, A., Hoover, R., Schoenberg, J.B., Gridley, G., Virgo, P.W. and Fraumeri, J.F. (1986) Carotenoid intake, vegetables, and the risk of lung cancer among white men in New Jersey. *American Journal of Epidemiology* 123, 1080–1093.

INDEX